Chemicals and Methods for Conservation and Restoration

Scrivener Publishing
100 Cummings Center, Suite 541J
Beverly, MA 01915-6106

Publishers at Scrivener
Martin Scrivener (martin@scrivenerpublishing.com)
Phillip Carmical (pcarmical@scrivenerpublishing.com)

Chemicals and Methods for Conservation and Restoration

Paintings, Textiles, Fossils, Wood, Stones, Metals, and Glass

Johannes Karl Fink

Scrivener Publishing

WILEY

This edition first published 2017 by John Wiley & Sons, Inc., 111 River Street, Hoboken, NJ 07030, USA and Scrivener Publishing LLC, 100 Cummings Center, Suite 541J, Beverly, MA 01915, USA

For more information about Scrivener publications please visit www.scrivenerpublishing.com.

Wiley Global Headquarters
111 River Street, Hoboken, NJ 07030, USA

For details of our global editorial offices, customer services, and more information about Wiley products visit us at www.wiley.com.

Library of Congress Cataloging-in-Publication Data
ISBN 978-1-119-41824-5

Cover images: Pixabay.com
Cover design by: Russell Richardson

Set in size of 11pt and Minion Pro by Exeter Premedia Services Private Ltd., Chennai, India

Contents

Preface

This book focuses on the chemicals used for conservation and restoration of various artefacts in artwork and archaeology, as well as special applications of these materials

Also the methods used, both methods for cleaning, conservation and restoration, as well as methods for the analysis of the state of the respective artifacts

The special issues covered concern:

- Oil paintings,
- Paper conservation,
- Textiles and dyes for them,
- Archaeological wood,
- Fossiles,
- Stones,
- Metals and metallic coins, and
- Glasses, including church windows.

The text focuses on the basic issues and also the literature of the past decade Beyond education, this book may serve the needs of conservators and specialists who have only a passing knowledge of these issues, but need to know more.

How to Use this Book

Utmost care has been taken to present reliable data Because of the vast variety of material presented here, however, the text cannot be complete in all aspects, and it is recommended that the reader study the original literature for more complete information.

Index

There are three indices: an index of acronyms, an index of chemicals, and a general index In the index of chemicals, compounds that occur extensively, e.g., *acetone*, are not included at every occurrence, but rather when they appear in an important context When a compound is found in a figure, the entry is marked in boldface letters in the chemical index.

Acknowledgements

I am indebted to our university librarians, Dr Christian Hasenhüttl, Dr. Johann Delanoy, Franz Jurek, Margit Keshmiri, Dolores Knabl, Friedrich Scheer, Christian Slamenik, Renate Tschabuschnig, and Elisabeth Groß for their support in literature acquisition In addition, many thanks to the head of my department, Professor Wolfgang Kern, for his interest and permission to prepare this text

I also want to express my gratitude to all the scientists who have carefully published their results concerning the topics dealt with herein This book could not have been otherwise compiled In particular, I would like to thank Dr Virág M Zsuzsanna for the provision of interesting details, which were very helpful for the preparation of this book

Last, but not least, I want to thank the publisher, Martin Scrivener, for his abiding interest and help in the preparation of the text In addition, my thanks go to Jean Markovic, who made the final copyedit with utmost care.

Johannes Fink
Leoben, 14th April 2017

1

Paintings

1.1 Cleaning

Historically, artists have protected oil painting surfaces with varnish. This is a system that allows the varnish to be brushed clean or even washed relatively frequently to remove accumulated surface dirt without exposing the paint to risk (1).

Unfortunately, mastic or other traditional soft-resin varnishes do not last indefinitely. After a few decades the varnish becomes yellow and brittle, losing transparency, and the cleaning process is transformed into the more challenging problem of removing the degraded varnish directly from the painting surface.

Even when new, a varnish may change the appearance of a painting. The varnish increases the transparency of any partly coated pigments or low refractive index medium, and also it imparts a new surface, which is frequently glossy. Mostly, artists have accepted such immediate changes in appearance for the future benefits of protection from dirt and from the risks of dirt removal.

By the eighteenth and nineteenth centuries, when state academies controlled much professional painting practice, the need for a varnish became important.

The concept of finish embodied many notions and became an unwritten contract of quality and reliability between academician and purchaser of art. It seems likely therefore that professional artists and their clients or patrons have always considered the application of varnish as a necessity of permanence and that artists have chosen to exploit its properties for both visual and practical benefit.

Many artists, through ignorance or untidy practice, continued painting up to exhibition deadlines and then immediately brushed varnish onto undried paint. A soft-resin varnish, such as mastic, was mixed into a paint to improve the short-term handling properties. Painting was even continued after varnishing. Adding a soft natural resin to oil paint remained popular into the middle of the 20th century (2).

Annual spring cleaning can be simply done by brushing or vacuuming dust from a varnish. However, washing with water is more effective and may need to be done only every decade or two decades. This procedure requires a wetting agent to ensure a good contact with the varnish surface and to trap dirt within the surface of the liquid.

Traditional recipes using potatoes and onions are well known (3). Saliva is still considered effective. Many other materials have been recommended, including borax and urine.

Conventional varnishes are most susceptible to UV radiation, air pollution, and moisture, and as the varnish ages, it becomes more polar and brittle and more soluble in aqueous mixtures. Aqueous methods for cleaning have been described in a monograph (4).

The varnish surface and, eventually, the body of the varnish disintegrate under the action of repetitive cleaning. Wax or poppy oil coatings can be applied to impregnate the varnish surface to extend its life, but opacity and yellowing may destroy its optical qualities (3).

Perhaps two generations will have passed since anyone saw the painting through a clear fresh varnish. The removal of a well-oxidized mastic varnish from a thoroughly dried oil film using spirits of wine has been carried out for centuries (5,6).

Alternatives to solvents have been favored by Wolbers (7). The cleaning of paint surfaces is done by using surface active agents in water-based systems. This can be effective in removing oxidized varnishes and oil varnishes as well as dirt. The formulations proposed by Wolbers have provided new tools to remove stubborn material more controllably (1).

1.1.1 Special Considerations

With the rapid developments in new cleaning techniques and analytical techniques it is important and necessary for the conservation community to constantly remind itself of the debate surrounding cleaning. In modern times, this debate began with the National Gallery of London *cleaning controversy* of 1947 (8). A scientific examination for art history and conservation has been published (9–11). The (surface) cleaning and the removal of varnishes are arguably the most controversial and invasive restoration interventions that a painting will undergo.

Doerner, already in 1921 published warnings about the damage that could be caused by solvents and cleaning (2, 8): The origins of the profession of painting restoration in France have been reviewed (12).

> There are countless cleaning materials, most of which are the secret of a particular conservator. One cannot believe all the possible types of materials which are applied to paintings. The strongest caustics, acids, and solvents are used without a second thought. Solutions with unknown composition, so-called secret solutions, are recommended to the public, as something anybody without any knowledge can use to clean pictures. Such cleaning methods are often too successful, right down to the ground layers. In those cases, the conservator covers up his sins by retouching.
> It is not uncommon that such locations appear cleaner to the unknowing public than the older version. Even to this day there are conservators who, in all seriousness, claim that they have cleaning materials which remove new paint but stop at the real, original layers. The only thing missing is that a bell should ring when the original paint layer is reached.

The use of balsams for cleaning paintings, in particular copaiba balsam, was fashionable until the end of the 19th century. However, the effect of this balsam was devastating and catastrophic, especially on oil paintings (13).

Copaiba balsam is a resin now known for its softening properties that remain active over a long period of time. An original paint layer treated with copaiba balsam is thus much more sensitive and subject to future damage than prior to the intervention. It is to be noted that commercial solutions such as Winsor and Newton Artists' Picture Cleaner still contain copaiba balsam (8).

1.1.2 Oxalate-Rich Surface Layers on Paintings

Oxalate salts have been the subject of extensive research as alteration products on calcareous substrates, e.g., stone and fresco. However, there has been relatively little notice concerning their occurrence on other objects such as easel paintings (14). The conservation of easel paintings has been reviewed (15).

An understanding of these materials is important since they can be responsible for significant changes in the surface appearance of artworks and the solubility of the matrices where the oxalates are formed.

Altered, oxalate-rich surface layers can causes substantial challenges for the visual interpretation of the painted surfaces.

Oxalate-containing layers or deposits have been reported on a variety of noncalcareous substrates, including glass (16,17), bronze (18–20), human remains such as mummy skin (21), and polychrome wood (22) and easel paintings (23–25).

The oxalate salts of calcium, whewellite (calcium oxalate monohydrate) and weddellite (calcium oxalate dihydrate), are those most commonly encountered on painted surfaces, although copper oxalates have also been identified in paint layers containing copper pigments.

Mostly these compounds have been found in deteriorated organic surface layers. Biological and chemical mechanisms have been proposed for the formation of oxalate films on artworks (26).

In the paintings studied in the Philadelphia Museum of Art, the oxalate minerals may likely derive from a gradual oxidative degradation of organic materials in the surface layers and their reaction with calcium-containing pigments or particulate dirt.

The resistance of the calcium oxalates to organic solvents and other cleaning agents presumably affects their enrichment on the surface (14).

1.1.3 *Leaching*

The cleaning of unvarnished paintings is one of the most critical issues. Several studies exist regarding different cleaning tools, such as gels, soaps, enzymes, ionic liquids, and foams, as well as various dry methods and lasers, but only a few have been performed on the risk associated with the use of water and organic solvents for the cleaning treatments in relation to the original paint binder (27).

The behavior of water gelling agents during cleaning treatments and the interaction of the following elements have been assessed: Water or organic solvents used for the removal of gel residues with the original lipid paint binder.

The study was conducted on a fragment of canvas painting from the 16th to 17th century of Soprintendenza per i Beni Storici, Artistici ed Etnoantropologici del Friuli Venezia Giulia, Udine, by means of Fourier transform infrared (FTIR) spectroscopy, gas chromatography (GC)/mass spectroscopy (MS), and scanning electron microscope (SEM) (27).

1.1.4 *Removal of Dirt*

The removal of dirt from an unvarnished paint surface may be very challenging, in particular, when the deposit is patchy and resilient; besides which, fragile unvarnished underbound paint surfaces are sensitive to aqueous solvents.

When the dissolved dirt may have impregnated the paint surface irreversibly, nonsolvent cleaning methods are necessary (28).

Dry surface cleaning uses a large range of specific materials like sponges, erasers, malleable materials, and microfiber cloths. However, these materials have not yet been fully integrated into the current practice of conservators. Only a few studies have focused on the use of dry cleaning materials in conservation. Most of the studies have focused on textile and paper conservation (29–32).

The testing methodology and results of dry cleaning materials on underbound and solvent-sensitive surfaces have been reviewed (28).

More than 20 cleaning materials used in conservation have been evaluated. This was based on preliminary cleaning tests on soiled and artificially aged oil paint surfaces. The materials are summarized in Table 1.1.

Table 1.1 Dry cleaning materials (28).

Type	Product name	Composition
Malleable material	Absorene	Starch, white spirit
Malleable material	Groom/stick	Isoprene, chalk
Eraser	Edding R10	PVC, DOP
Eraser	Pentel ZF 11	PVC, DOP, etc.
Eraser	Bic Galet	Vegetable oil
Cloth	Yellow microfiber	PET, PA
Sponge	Smoke sponge	Isoprene rubber
Sponge	Akapad white	Styrene butadiene rubber
Makeup sponge	Etos	Isoprene rubber
Makeup sponge	Hema	Styrene butadiene rubber
Makeup sponge	QVS	Poly(urethane)
Gum powder	Draft clean powder	Styrene butadiene rubber

Aging procedures were performed for 4–6 weeks at temperatures of 50–60°C with variations of relative humidity from 27% to 80% every 6 *h*. Light aging was done with fluorescent tubes (10,000 *Lux*) for approximately 600 *h* at a temperature of 23°C and a relative humidity of 44%. This is equivalent to 11.5 *y* of aging under museum conditions.

The first series of tests were performed on a naturally aged 30 *y* old monochrome oil painting on canvas. The second series of tests were performed on water sensitive cadmium red, cadmium yellow, and ultramarine blue tube oil paints. The third series of tests were performed on Gouache samples.

Dry cleaning tests were performed under ambient temperature and humidity. After each test, the paint samples were brushed and vacuum treated. The test results were observed visually, then using light microscopy, followed by electron microscopy.

The test results indicated that the Akapad white and makeup sponges were the least abrasive polishing materials. Both materials are very efficient for the removal of embedded and resilient dirt. In contrast, eraser-type materials proved to be the most harmful materials. Here, chemical residues, i.e., the plasticizers, were detected in the paints. This is a special issue, since plasticisers can soften the paint surface, leaving it more sensitive to dust and vulnerable to abrasion or polishing. On the other hand, Groom/stick and Absorene left a film deposit or particulate residue on both well-bound and porous paint layers. This deposit may harden and embed into the paint layer in the course of aging. In summary, makeup sponges proved to be the most efficient and the safest materials (28).

1.1.5 *Effects of Organic Solvents*

Several technical studies of the effects of solvents on oil paints in the context of removal of varnish from paintings have been reviewed (33). Also, the historical background of technical studies of cleaning and the various effects of solvents on oil paints have been discussed. These include (33):

- Swelling and softening of the paint binder, which can contribute to the vulnerability of paints to pigment loss during cleaning,
- Solvent diffusion and retention, and
- Leaching, i.e., the extraction of soluble organic compounds from the paint.

The methodological issues in cleaning studies have been discussed, particularly the relationship between studies on model reference paint films and realistic, clinical studies of actual cleaning operations, also considering the related issue of aging of oil paints (33).

1.1.5.1 *Solubility Parameters*

A number of systems for the specification of solubility properties have found currency in the field of conservation (34). The theoretical foundations of various extant solubility parameter schemes have been critically reviewed in the context of the cleaning of paintings with organic solvents.

Recent advances in solvency specification are discussed, and comprehensive tables of solubility parameter data have been compiled from various sources. One recently developed scheme is that of Snyder and co-workers. This scheme provided the foundation for the proposal of a new composite solubility parameter scheme with potential applications for aiding solvent selection in cleaning and for describing the swelling response of paints to solvents.

It has been proposed that this scheme provides the foundation for an improved understanding of the internal cohesive chemistry of paint films (34). The nature of solubility parameters have been extensively reviewed (35).

The Teas f_d solubility parameter is an indicator for solubility (36). Teas solubility parameters are normalized Hansen solubility parameters. The solubility of coatings has been detailed (37).

Values for maximal swelling of burnt umber linseed oil films, aged 12 days at 80°C for various solvents, are collected in Table 1.2. Some of the compounds are shown in Figure 1.1.

Table 1.2 Values for maximal swelling of burnt umber linseed oil films (36).

Solvent	Teas f_d	Paint film Thickness/μm	Average area Swelling
Perfluorodecalin	100	230	0.7
i-Octane	100	330	−4.75
White spirits	90	230	7.52
Tetrachloromethane	85	330	1.5
Ethylbenzene	87	320	9.7
Dibutyl ether	70	230	10.9
Dioxane	67	220	23.5
Amyl acetate	62	370	11.6
Cyclohexanone	55	220	25.6
Dichloromethane	59		38.8
Butanone	53	310	20.3
IMS/iso-octane	68	340	7.3
Acetone	47	330	19.8
N-Methylpyrrolidone	48	300	34.7
t-Butanol	44	230	6.9
DMSO	41	230	22.2
Propan-2-ol	38	320	5.1
Butan-1-ol	43	330	6.8
Methoxypropanol	42	300	14.3
Ethanol	36	300	15.6
IMS	36	360	9.5
Acetone/water 1:1	32.5	230	18.5
Methanol	30	310	17.4
Trifluoroethanol		220	23.0
Triethanolamine pH 9.7	n/a	230	37.9
Ammonium hydroxide pH 11.2	n/a	230	52.5

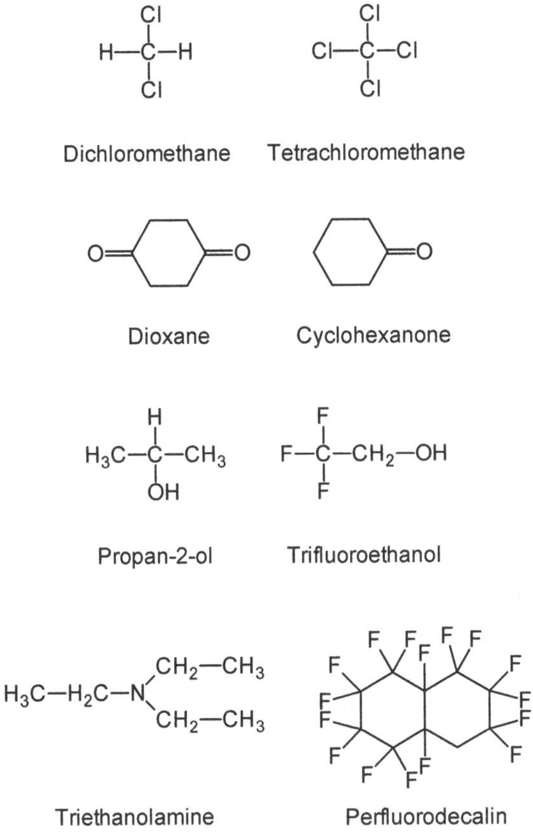

Figure 1.1 Solvents for swelling tests.

Further, solvents used for resin solubility testing and their Teas fractional solubility parameters have been detailed (38).

It has been stated that Teas charts have come under fire for a number of simplifications, shortcomings, and *fudge factors*. Two of the most cogent attacks have been summarized (39,40)

In short, the Teas system can be criticized for overemphasizing the dispersion forces, neglecting ionic and acid-base interactions, rejecting the overall differences in the magnitude of cohesive energy densities, and assuming solvent and solute randomness (38).

The swelling responses of two oil paint films as a consequence of immersion in solvents of various kinds have been elucidated (41). Two test paint films with the same original formulation are based on proprietary artists' oil colors containing yellow ocher and flake white pigment bound in linseed oil. One was aged by exposure to high light dosage, and the other was unexposed. Lateral, inplane swelling of the paint films during immersion in solvent was determined by a simple microscopical method using computer-based image analysis.

Results have been reported for the swelling of both paint films in more than 55 common solvents and 14 binary solvent mixtures containing ethanol. The data have been presented as swelling curves of percentage change in area as a function of time and as plots of the degree of maximal swelling as a function of selected solvency indicators. The results have been discussed in comparison with existing data on the swelling of oil films by organic solvents and in relation to the implications for the cleaning of oil-based paints (41).

In research and in actual conservation practice, the conservators have to choose adequate methodologies for carrying out treatments successfully, while respecting the integrity of artworks (42). In particular, the conservators must be able to choose appropriate conservation materials and methods.

Solvents are widely used in cleaning, but solubility issues are also of high importance in consolidation treatments as well as in protective coating applications.

The potential of Hansen solubility parameters for reliable use in the field of artwork conservation has been checked (42). An effort was made to develop an efficient methodology for critical solvent selection.

For this purpose, two different methods were used for the estimation of various artwork conservation materials. A group-contribution method, based on the chemical composition of materials, was applied for the prediction of Hansen solubility parameters of egg yolk, pine resin and seven red organic colorants (Mexican, Polish and Armenian cochineal, kermes, madder, lac dye and dragon's blood), traditionally used in paintings, textiles and illuminated manuscripts.

Additionally, an experimental setup was used for testing the solubility of the commercial products of synthetic conservation materials, Primal AC-532K, Beva gel 371 a and b, as well as a commercial matt varnish made of dammar and wax. The direct use of Hansen solubility parameters and the relative energy difference between various materials made it possible to carry out ad hoc virtual solubility tests that may apply to real and complex systems such as cultural heritage artworks (42).

1.1.6 *Cavitation Energy for Solvent Mixtures*

The use of solvent mixtures for surface cleaning in restoration and conservation is widespread. However, there is a lack of knowledge on the true consequences of such a treatment (43).

Azeotropic solvent mixtures have been proposed. It is well known that binary solvent mixtures behave nonideally. This means that the properties of the mixture are neither proportional nor related to the mixing ratio.

The solubility of a material is controlled by the solubilization of the solute and the molecular stabilization of the solute within the liquid phase. There is a difference in the behavior between a solvent mixture and either of the pure solvents as both their solvation properties and their cavitation energy vary significantly.

Solvation relates to the intermolecular forces between the solvent and the solute. A selective solvation may arise from a greater affinity of one component of the solvent mixture to the macromolecules or other components of the paint film (44).

Of particular interest in practice is the cosolvation effect, where each solvent exhibits a selective affinity to one type of structural element. This may lead to an increased solubility of a bistructural material, such as alkyd paints, which contain a phthalic acid polyester backbone in addition to fatty acid substituents. Often, the energy of cavitation is ignored in the considerations.

The free energy of solubilization ΔG_m is (45):

$$\Delta G_m = \Delta H_m - T\Delta S_m \tag{1.1}$$

In a dissolution process, the free energy of mixing ΔG_m must become lower in the course of solubilization. The enthalpy of mixing ΔH_m requires similar intermolecular solvent-solvent and solvent-solute forces for a successful action and is mostly positive and small. Therefore, the entropy of mixing ΔS_m at a given temperature T is of relevance.

The change in entropy in the course of mixing is mainly dependent on the strength of the intermolecular interaction within the liquid because the liquid cohesion has to be overcome to form a cavity in the liquid to incorporate the solute (46).

The cavity formation can be energetically described by the cohesive energy of the liquid. This can be qualified by the Hildebrand parameter δ_H^2. This parameter controls the entropy of the dissolution process.

In the process of dissolution both endothermic and exothermic steps occur. The exothermic step is an enthalpic process which can be described by the intermolecular interaction between solute and solvent. These interactions may be dispersive, aprotic, or protic.

In a study, the swelling capacity upon immersion of paint films in organic solvent compositions was used to quantify the solvation effects on the binder matrix.

The experiments were done using six solvents, i.e., *n*-hexane, toluene, chloroform, diethyl ether, acetone, and ethanol, as well as binary mixtures.

Extracts of 2 *g* of paint sample in 50 *ml* of solvent were gravimetrically quantified and also characterized using FTIR, direct temperature resolved mass spectrometry, and GC MS. The FTIR studies suggested that the increasing polarity of the solvent mixture results in increased leaching of polar oily components. At swelling levels where changes in volume exceed 7% by volume a massive increase of triglycerides in the leached materials was found.

The swelling data reveal almost equivalent swelling anomalies within oil and alkyd paints. In extreme cases the swelling volume may reach several times the ideal value. This effect is not influenced by the liquid-solid interactions but is caused by liquid-liquid interactions. It has been found that the larger the difference in polarity is between the mixed solvents, the greater the observed deviation is from the ideal behavior.

On the other hand, in apolar mixtures the deviation from the ideal behavior is small. In contrast, mixtures that contain a polar solvent may exhibit strong anomalies in swelling behavior. Thus, ethanol-containing mixtures induce very strong swelling anomalies in oil and alkyd paints, with an increase in volume of up to 200%. This effect is particularly pronounced in ethanol mixtures that form azeotropes.

The swelling anomalies correlate with a change in the boiling point (47). The swelling data have been documented in much detail

(43). The properties of solubilization and the swelling capacity of solvent mixtures are directly relevant to the extraction of low molecular compounds in paintings.

1.1.7 Hydrogels Based on Semi-Interpenetrating Networks

Water-based detergent systems offer several advantages over organic solvents for the cleaning of cultural heritage artifacts in terms of selectivity and gentle removal of grime materials or aged varnish, which are known to alter the readability of the painting (48).

Unfortunately, easel paintings show specific characteristics that make the usage of water-based systems invasive. The interaction of water with wood or canvas support favors mechanical stresses between the substrate and the paint layers, leading to the detachment of the pictorial layer.

In order to avoid painting loss and to ensure a layer-by-layer control of grime removal, water-based cleaning systems have been confined to innovative chemical hydrogels, specifically designed for cleaning water-sensitive cultural heritage artifacts.

The hydrogels are based on semi-interpenetrating chemical poly (2-hydroxyethyl methacrylate)/poly(vinylpyrrolidone) networks with a suitable hydrophilicity, water retention properties, and sufficient mechanical strength to avoid residues after the cleaning treatment. The monomeric compounds are shown in Figure 1.2.

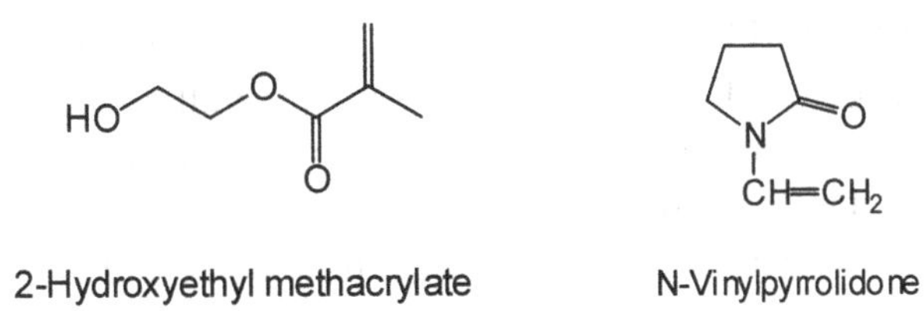

Figure 1.2 Hydrogel monomers.

The water retention and release properties have been studied by quantifying the amount of free and bound water using differential scanning calorimetry (DSC). The mesoporosity was obtained from SEM. The microstructure was assessed using small angle X-ray scattering. The efficiency and versatility of the hydrogels in confining and modulating the properties of cleaning systems was shown in a case study (48).

1.1.8 *Organogels*

Organogels have been described as cleaning tools for painted surfaces (49). These combine the most attractive features of cleaning liquids and normal gels while diminishing the deleterious characteristics of both.

The latent gellant, poly(ether imide) (PEI), reacts with CO_2 at room temperature in organic solutions to produce an ammonium carbamate form PEI CO_2. Ammonium carbamate is a salt that is formed by the reaction of ammonia with carbon dioxide or carbamic acid. The compound is shown in Figure 1.3.

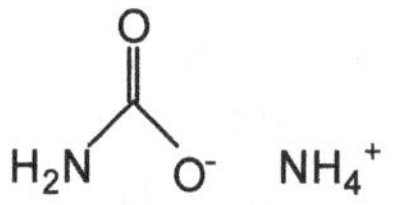

Figure 1.3 Ammonium carbamate.

The charged moieties turn into three-dimensional polymer networks that immobilize the liquids as gels.

The properties of the initial solution can be reestablished by the addition of a small amount of a weak acid. The acid displaces the CO_2 molecules and the PEI chains become positively charged.

Contact angle and FTIR measurements as well as visual comparisons of the surfaces before and after application of the gels have been done. The visual changes are substantiated by rheological measurements. The results indicated that these gels are valuable cleaning tools for painted surfaces of historical and artistic interest.

In particular, the PEI CO_2-based organogels are highly effective in removing certain surface patinas from painted supports. A surface layer of dammar was completely removed from a test canvas with oil paint, an aged painting from the 19th century, and a 15th century oil-on-wood panel attributed to Mariotto di Cristoforo. In addition, a surface acrylic polymeric resin which was used in a restoration performed during the 1960s was successfully removed from Renaissance wall paintings decorating the Santa Maria della Scala Sacristy in Siena, Italy.

The isothermally rheoreversible gel represents a new, highly versatile, and very efficient method for removing aged surface patinas from works of art (49).

1.1.9 Microemulsions and Micellar Solutions

Nanoscience and nanotechnology have evolved as revolutionizing materials science. These materials are creating new approaches to conservation science, leading to new methodologies that can revert the degradation processes of works of art. The most advanced current methodologies are (50):

1. The use of water-based micelles and microemulsions, neat or combined with gels for the removal of accidental contaminants and polymers used in past restorations, and
2. The application of calcium hydroxide nanoparticles for the consolidation of works of art.

The combination of different nanotechnologies allows a conservator to provide, in each restoration step, interventions respectful of the physicochemical characteristics of the materials used by the previous artists. The methods using nanoscience are continuously growing (50).

The synthesis and preparation of colloidal systems tailored to the consolidation and protection of wall paintings, plasters and stones have been detailed. Also, two case studies, widely representative of typical consolidation problems, have been discussed, namely the preservation of wall paintings belonging to a Mesoamerican archaeological site and the consolidation of two Italian Renaissance buildings (51).

1.1.10 Acrylic Paintings

The Modular Cleaning Program has been developed by the Getty Conservation Institute, California State University, the University of Delaware, and the Conservation Division of the Winterthur Museum.

The target of the project was to systematically investigate the problem of residues left on surfaces after cleaning with a system that contained nonvolatile components.

The Modular Cleaning Program should help the conservator to achieve precision in designing a cleaning system. It does not direct the conservator on how to clean a surface, but rather organizes the cleaning options in a logical and expandable way.

The Modular Cleaning Program models the physical and chemical interaction. It has 19 internal databases, but the conservator interacts directly with only a few, the components database. The components database includes (52):

- Components database,
- Repository of physical constants,
- Chemical names;
- Chemical Abstracts Service registry number,
- Molecular weight,
- Physical form,
- Density,
- Acid dissociation constants for weak acids and bases,
- Formation constants for chelating agents,
- Hydrophile-lipophile balance number,
- Critical micelle concentration,
- Aggregation number,
- Cloud point for surfactants,
- Boiling point,
- Hildebrand, Hansen, and Teas solubility parameters,
- Dipole moment,
- Index of refraction,
- Dielectric constant, and
- Molar volume for solvents.

There is also a separate database of binary and ternary azeotropes for solvent mixtures.

Thus, the modular cleaning program accesses the relevant physical constants as it calculates suitable cleaning formulations in the solutions database and another of the databases actively used by the conservator (52).

Aqueous solution calculations are based on the appropriate physical constants from the components database and user-specified data.

For pH buffer and chelating agent solutions, acid dissociation constants and user specifications of counterions, cleaning solution concentrations, and pH values are used to calculate the needed concentrations of ionic and molecular species in solution and the amount of the counterion necessary to set the solution to the specified pH.

In addition, solvent mixtures are modelled in the three-dimensional Hansen space, but Teas and Hildebrand parameters are also calculated. The solvent phase of Carbopol®-based solvent gels and Pemulen®-based emulsions are modeled as simple solvent mixtures. The gel phase of solvent gels is based on stoichiometric relationships between the Carbopol, the user-specified organic amine or amines used to neutralize the Carbopol, and the inferred stoichiometric amount of water necessary to form a gel. The aqueous phase of Pemulen-based emulsions is modeled as an aqueous cleaning solution thickened with Pemulen to which a solvent or solvent mixture is added (52). Pemulen polymers are high molecular weight, crosslinked copolymers of acrylic acid and C_{10}–C_{30} alkyl acrylate. Acrylic acid is shown in Figure 1.4.

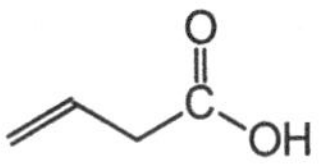

Figure 1.4 Acrylic acid.

Information on the background and use of the modular cleaning program has been detailed (53). The modular cleaning program software is freely available to professional conservators.

1.1.11 *Acrylic Emulsion Paintings*

Cleaning acrylic emulsion paintings is challenging because of the material properties of the paint films, in particular their solubility (54). The effect of aqueous treatments on acrylic paints was investigated. Several paint manufacturers were asked for their recommendations for cleaning these paintings and conservators were also asked to comment on the damage observed in them and on the treatments applied.

Their responses showed that aqueous cleaning treatments are used, despite the associated risks, and that more technical information is needed about the effects of cleaning. The changes in the physical and mechanical properties of an aged cobalt blue paint were evaluated as a result of the exposure to aqueous cleaning solutions. The results suggested that a short immersion in these solutions caused a drop in most of the mechanical properties. Interestingly, longer immersions did not cause a drop in the mechanical properties.

The drop after the short immersions seemed to be mostly due to the great increase in the thickness of the paint films (54).

1.1.12 *Complications in the Cleaning of Acrylic Paint Surfaces*

Acrylic paints are in many ways remarkably durable. However, the surfaces are rather fragile. This fragility, coupled with the tendency for the paintings to be quite large, leaves them vulnerable to damages from handling. Fingermarks and scuffs can be disfiguring and difficult to ameliorate (52).

In addition, the gradual accumulation of surface grime is of general concern (55). It has been shown that even in a presumably benign museum environment the slow accumulation of surface grime will lead to a discernible visual alteration of a paint surface after approximately 50 years.

Research has been done to identify some of the special properties of acrylic paints (56,57).

The examination of acrylic paint formulations has revealed the variables that must be considered when devising a cleaning strategy for these paint surfaces (7,58).

Actually, the base of an acrylic emulsion paint system is the acrylic emulsion itself. It is created from water, the acrylic monomer, or monomers, and some materials to stabilize the monomers into micelles, and also to control the pH, initiate the polymerization reaction, and prevent foaming.

Additional ingredients, such as thickeners and coalescing agents, transform it into a successful paint binder.

Furthermore, pigments and extenders must be mixed with water, dispersing agents, and other components that are then added to the composition of the acrylic emulsion paint.

All of these materials form a very complicated and interdependent system compared to any traditional paint system. So an attempt can be made to predict the potential sensitivity of the paint on the basis of the sensitivities and properties of each of these many components.

These potential sensitivities can be categorized into some broad classes related to the effects of (52):

- pH,
- Ionic strength, and
- Surfactant migration.

Also, it is important to think about the physical structure of the acrylic paint film.

Acrylic emulsion paints dry in a different way than other paint systems. The initial drying occurs by evaporation of much of the water, leaving the pigments and spherical polymer-containing micelles to form a close-packed matrix. In the next stage of drying the water held by capillary forces between the polymer spheres and pigment particles is evaporated.

Coalescing agents and the nature of the polymer itself allow the spheres to diffuse into adjacent spheres, forming a film (59).

The resulting structure of the film will be between the solid membrane of a fully coalesced film and an inhomogeneous structure with pores, irregularities, and imperfections.

So, the physical structure of the acrylic film has particular implications when considering surfactant migration and the ionic strength of the cleaning systems (52).

1.1.13 *Poly(vinyl acetate) Paints*

Synthetic paints are mostly a complex mixture of additives that are likely to influence the response of the paint to different treatment methods. Furthermore, their high sensitivity to most organic solvents narrows the choice of products that can be used.

The knowledge that has been obtained during the cleaning of traditional oil or egg tempera paintings cannot be correlated to contemporary paintings. Therefore, additional research concerning the chemistry and properties of synthetic paints and their response to the conservation practice is needed.

Preliminary results on the effects on test films of poly(vinyl acetate) (PVAc) paint have been published (60). These studies were done for cleaning paintings by Julião Sarmento. This artist is one of the most prominent contemporary Portuguese artists who has made wide use of PVAc paints. In the 1990s he started a series of paintings, commonly called the *White Paintings*. These paintings already require a surface cleaning because of a combination of their monochromatic white backgrounds and their affinity for dirt pickup.

Samples of a PVAc emulsion paint containing titanium dioxide in the rutile form were artificially aged using a xenon arc light source for 3250 *h*. In the course of the irradiation, the mock-ups were exposed to dirt particles, i.e., atmospheric particles. Commonly used cleaning methods were subsequently tested:

1. Water,
2. Water and a nonionic surfactant, i.e., Brij700S, and
3. White spirit.

Brij is a poly(ethylene glycol) octadecyl ether compound. The use of a soft eraser (Akapad white) was also examined to reproduce a real case of cleaning a work by Sarmento. White spirit was included since aliphatic mineral spirits are known to have a minimal effect on acrylic paints (57).

Since it is known that most of the material lixiviated from a paint sample occurs in the first few minutes, the effect of a 5 *min* immersion was evaluated.

Artificial aging studies indicated that the loss of the paint plasticizer, diisobutyl phthalate, is the major alteration between aged and unaged samples. Diisobutyl phthalate is shown in Figure 1.5.

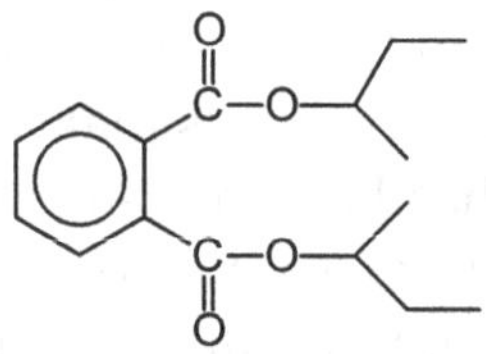

Figure 1.5 Diisobutyl phthalate.

The FTIR spectra from the artificially aged control sample showed no significant signs of surface enrichment with diisobutyl phthalate. Therefore, it was expected that immersion could only remove negligible quantities of plasticizer still present on the test films. Only minor changes were observed in the FTIR spectra that can be correlated with the extraction of this additive. Moreover, diisobutyl phthalate was not detected among the residues remaining after immersion (60).

Because of the irregular surface topography and the affinity for the dirt that is retained by the paint surface, the immersion of the samples was not efficient for the removal of dirt. This was confirmed by optical microscopy and also by colorimetry.

Pure water did not induce relevant alterations concerning the morphology of the white paint. However, with an aqueous solution of Brij700S, the smoothness of the latex particles is found to be lost. Here, the polymer surface displays a more irregular texture, which could be the result of surfactant residues left on the paint.

The cleaning with an eraser resulted in both particles and/or eraser additives left on the surface.

A physical disruption of the latex paint was only detected at the nanoscale. However, it could result in a flattening of the paint surface that is contrary to the distinctive pigment agglomerates and filler texture typical of these paintings (60).

1.1.14 Surface Cleaning

The developments of cleaning methods for paintings in Italy over the course of the last two decades have been described (61). In Italy the restoration has been mostly practiced by private restorers.

Public laboratories, such as those that belong to museums, still represent only a small minority. The research for less toxic materials and a more reasonable and cautious use of the traditional materials is still a concern for restorers.

In the second half of the 20th century, the cleaning of easel paintings and polychrome objects in Italy was mainly performed with neutral organic solvents and acid and alkaline reactants, sometimes mixed together and/or with water. One of the most commonly used mixtures was obtained by mixing water with an alkaline substance in variable proportions.

Testing was performed with mixtures of different solvents that were ready in the studio for use and were considered efficient enough to remove ingrained surface dust rather than varnishes, unknown coatings on paintings, old retouchings, and old fillings.

Some common mixtures used for restoration are collected in Table 1.3 and some are shown in Figure 1.6.

Table 1.3 Common mixtures used for restoration (61).

Components	Ratio
Water + ammonia	1:1
Water + acetone + ethanol	1:1:1
Water + ammonia + ethanol + acetone	1:1:1:1
Water + butylamine	1:1
Water + butylamine + DMF	1:1:1
DMF + amyl acetate	1:1
DMF + amyl acetate + nitro thinner	1:1:1
DMF + synthetic thinner + xylol + acetone	3.5:1.5:1:1

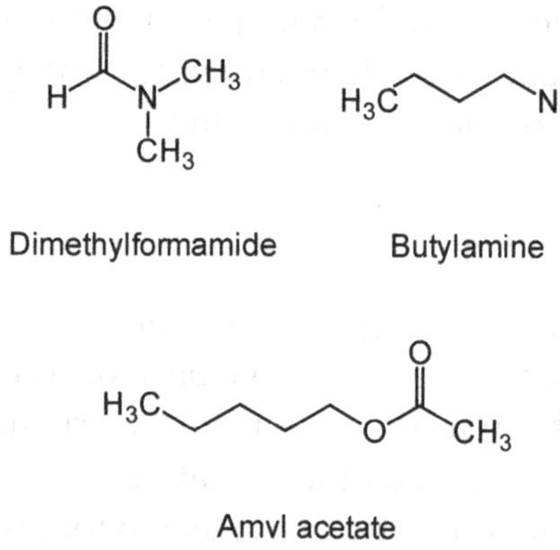

Figure 1.6 Compounds used for restoration (61).

1.1.14.1 *Surface Cleaning with Aqueous Foams*

Foam is a two-phase system consisting of a liquid and a gas phase that generally tends to separate into its two phases. Thus it is a thermodynamically unstable system. Under the influence of gravitational forces the liquid drains out of foam (62).

For restoration purposes, it is important to develop foams that have a high long-term stability and a slow, predictable release of the liquid phase. Without these properties, foams are useless for conservators.

The formation of more stable foam structures can be achieved by foaming agents. Foaming agents are surface active materials that reduce the surface tension of the liquid.

Anionic and nonionic surfactants and additional foam stabilizing polymers, i.e., cellulose ether, were tested at different concentrations to find the optimal formulations for foam generation. The materials are shown in Table 1.4.

Table 1.4 Surfactants and polymers for foam and gel formulations (63).

Ionic class	Chemical name	Product name	Producer
Nonionic	Fatty alcohol ethoxylate	Marlipal 618/25	Sasol
Nonionic	Poly(ethoxysorbitan)	Tween 20	ICI
Nonionic	Alkyl polyglucoside	Plantacare 818 UP	Cognis
Anionic	Sodium dodecylsulfate	SDS	ROTH
Anionic	Fatty alcohol ether sulfate	Texapon NSO	Cognis
Nonionic	Hydroxypropyl cellulose	Klucel H	Hercules
Nonionic	Methylcellulose	Methocel A40M	DOW

The foams were produced in a foam-generating apparatus with a rotating stirrer. To test the cleaning performance, an alkyd-based paint layer, Winsor & Newton Griffin Alkyd light cadmium yellow, was artificially contaminated with a spray-applied dirt.

The test results indicated that the use of a foam can help to minimize the draining and migrating of fluids in comparison to the use of a cotton swab. However, the tested foams released more water than the tested gel.

It has been concluded that foams have a general cleaning effect that is similar to the traditional cleaning methods. The best cleaning results were found to be the foam system containing alkyl polyglucoside and cellulose ether (63). In summary, a benefit of foam cleaning is the reduction of the mechanical stress during the removal of the foam by vacuuming.

1.1.15 Foxing Stain Removal

Foxing is a pattern of spotting or speckling stains that mar many archival works found in old books, vintage paper, or cloth. The pattern is usually brown or yellowish brown in tone and often more or less circular in shape. Its cause is not fully understood but generally it is believed to be a slow process caused by fungal or mold microorganisms enabled by impurities in paper and storage conditions that are damp and warm enough to facilitate the process.

Filamentous fungi are known to damage and destroy paper and celluloid items in two principal ways. First, they utilize the paper cellulose as a carbon source, weakening and eventually destroying the paper fibers. Fungi also live on the trace metals found in paper or often in the inks on the paper. Four strains of fungi are commonly found to cause foxing and each fungi is characterized by the production of different colored stains.

The fungus *Alternaria solani* produces a dense black stain, *Fusarium oxysporum*, a pinkish stain, *Penicillium notatum*, a light green stain and *Chaetomium globosum*, a brownish grey stain. These stains can sometimes be extracted with harsh solvents but there are few effective solvents that do not dissolve the ink or damage the paper fibers and many stains resist solvent extraction effectively.

The treatment for foxing is difficult at best and often simply impractical. The development of new solvent systems is time consuming and requires a great deal of trial and error, since the chemical structure of the pigment stains is not generally known. Mechanical stain removal is also problematic in that it is not selective between ink and stain; it often produces abrasion of the paper fibers, markedly deteriorating the paper.

The use of unsafe and damaging treatments to remove foxing stains is well established in the art of restoration and conservation. Typical topical chemicals used in the art are summarized in Table 1.5. Some chemicals are shown in Figure 1.7.

All of these dangerous chemicals listed in Table 1.5 can seriously degrade the cellulose in the paper and may lead to wrinkling of the pages or bleeding of text or illustrations.

These chemicals also introduce acidic residue salts that will contribute to additional damage over time and are difficult to use and quite expensive to purchase.

Table 1.5 Common chemicals used to remove foxing stains (64).

Chemical	Chemical
Ethylene oxide	Sodium hychloride
Chloramine-T	Calcium hypochlorite
Chloramine gas	Hydrogen peroxide
Sodium borohydride	Chlorine dioxide
Chlorohexidine gluconate	Magnesium oxide
Potassium permanganate	Lithium aluminum hydride

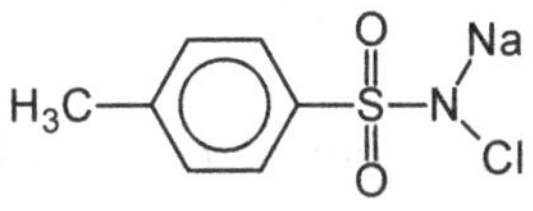

Chloramine-T

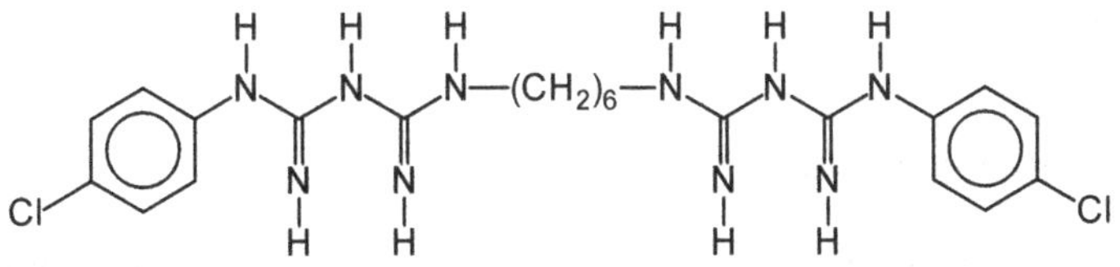

Chlorohexidine

Figure 1.7 Chemicals to remove foxing stains.

The difficulties explained above can be avoided using the following method (64):

1. Soak the paper or celluloid item completely in a warm bath of distilled water,
2. Remove the paper or celluloid item and blot all excess water,
3. Soak the paper or celluloid item completely in a warm bath of household bleach, less than 5% concentration,
4. Soak and rinse the paper or celluloid item completely with warm distilled water,
5. Remove the paper or celluloid item and blot all excess water,
6. Soak the paper or celluloid item completely in a warm bath of vinegar, no less than 75% concentration,
7. Soak and rinse the paper or celluloid item completely in a bath of cold distilled water,
8. Remove the paper or celluloid item and blot all excess water,
9. Soak the paper or celluloid item completely in a bath of sodium bicarbonate (baking soda) and distilled water,
10. Soak and rinse the paper or celluloid item completely with distilled room temperature water,
11. Remove the paper or celluloid item and blot all excess water,
12. Soak the paper or celluloid item completely in a bath of glue size,
13. Soak and rinse the paper or celluloid item completely in a bath of cold distilled water,
14. Remove the paper or celluloid item and blot all excess water,
15. Air dry the paper or celluloid item without the use of a heat blower.

1.1.16 Vacuum Techniques

In the early 1990s, the Schlürfer was developed for dirt removal from sensitive modern art paintings by the German conservator Winfried Heiber (65). The apparatus is shown in Figure 1.8.

The equipment consists of a vacuum pump equipped with a regulator and a gauge. The pump is connected to a gas-washing bottle, optionally fitted with a filter disc to reduce the size of air bubbles, by a flexible hose. Another flexible hose is connected to the other side of the gas-washing bottle and has a soft moistened suction poly(vinyl

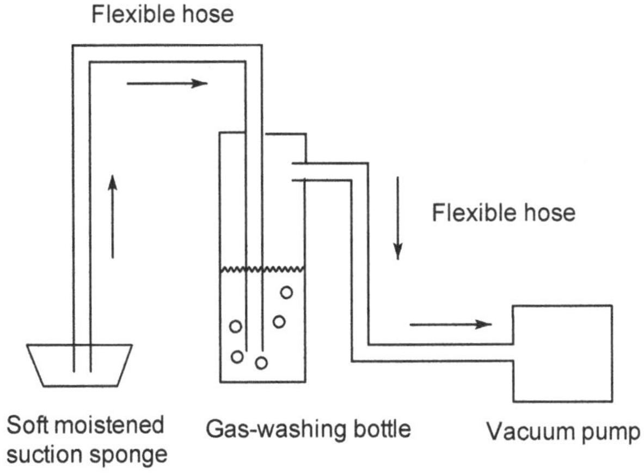

Figure 1.8 Schlürfer apparatus (65).

alcohol) (PVA) sponge at its end. The sponge is a highly absorptive, microporous sponge.

Because of the suction of the vacuum pump, the solubilized dirt is vacuumed through the soft wet sponge during the cleaning process. Both dirt and solvent are collected in the water-filled gas-washing bottle. A cotton filter pad in front of the outlet is generally recommended to protect the pump.

Because the suction power of the Schlürfer is much higher than that of sponges, cotton fibers, microfiber tissues, and others, the water can be removed rapidly from the surface of the painting.

A water treatment with the Schlürfer could reduce the mechanical impact to such an extent that no paint damage or negative effects could be observed afterwards (65).

1.1.17 Laser Cleaning Removal

The removal of past and unsuccessful conservation treatments from the surfaces of works of art is a difficult task that requires exceptional control and selectivity.

Many polymers (including acrylic and vinyl-based proprietary materials such as Paraloid® B-72, Primal, Plextol, Plexisol, Elvacite, and Laropal) have been applied to consolidate or varnish cultural heritage objects, including stones (66–68) and paintings (69–71).

However, polymer degradation and undesired surface effects are induced by the presence of polymers. For example, glossiness, color saturation, opaque and uneven discolorations or blooming can compromise the legibility and appearance of works of art (69,72).

The use of solvents, either aromatic organic compounds, or gels and emulsions, have been proposed for removing degraded polymer treatments which may obscure the surface of wall paintings (33,73,74).

However, the control of solvent penetration and clearance of materials after cleaning may be problematic. For example, emulsions and solvents may penetrate deeply into the bulk of the treated surface and may cause migration of organic materials, which can be extensive in porous structures found in wall paintings and stone substrates. Furthermore, health risks and practical issues, i.e., in the case of extended surfaces, may prevent the application of such solvent-based cleaning methods to wall paintings. These are often found on external surfaces exposed to ambient conditions. Similarly, the use of enzymes and bioremediation methodologies (75), which may require high temperature and buffer solutions, have been suggested for the removal of polymeric consolidants, but their application may be limited due to feasibility and control.

The potential of using lasers to remove synthetic resins used as consolidation materials on painted surfaces has been the objective of a multidisciplinary research scheme motivated by the large-scale conservation project at the Monumental Cemetery of Pisa, Italy (76).

An explorative study has been done to investigate the potential of laser cleaning for the removal of past consolidation treatments on real fragments taken from this monument as well as on specially prepared model samples; a number of effects were observed such as microcracking of the polymer film, loss of transparency (mainly in the *ns* regime) and a halo of redeposited material around the irradiated areas (mostly in the *fs* regime).

Some stereo micrographs of laser irradiation spots on model wall painting fragments are shown in Figure 1.9.

Although the laser-assisted removal of aged varnishes from painted surfaces has been extensively examined (77–80), laser cleaning of commonly found consolidants has not been addressed to the same extent (81).

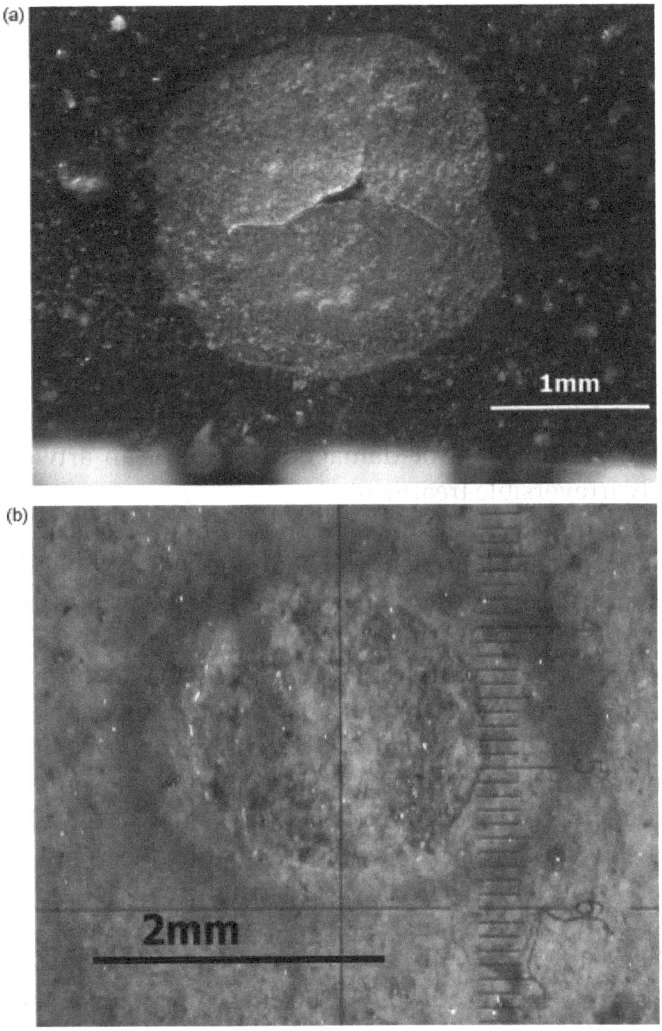

Figure 1.9 Stereo micrographs of laser irradiation spots on model wall painting fragments covered with a thin film of polymeric coating upon laser irradiation with a KrF excimer at 248 *nm* of (a) *ns* and (b) *fs* pulse duration. Reproduced from (76) with permission from Elsevier.

UV cleaning at 248 *nm* of polymerized, crosslinked, and discolored protective varnish layers that were originally applied to saturate the colors, so as to change the appearance of a painting and to protect the painting surface, may be used to achieve a clean and controlled material removal. In this way, a minimal, if any, chemical alteration to the underlying surface is done.

Since in most cases painting surfaces are sensitive to environmental conditions, such as light, heat, or humidity, and photothermal effects may cause a discoloration of the original paint or the pigments, a partial laser cleaning of the outermost degraded varnish layer, the combined use of mild solvent and solvent gel cleaning, and the reapplication of a new varnish coating are key steps in conservation and restoration.

In contrast to the use of varnish, the application of consolidation materials to paintings usually aims to fix, stabilize and protect the fragile, deteriorated or damaged surface of an artwork in an effectively irreversible treatment.

Several synthetic resins have been used to protect sensitive surfaces, to consolidate degraded wall paintings and to strengthen fragile wooden structures, to adhere various layers as well as to enhance the properties of painting media.

It has been common to apply a protective layer of polymer to the surface of exterior wall paintings, although the potential problems associated with temperature fluctuations and insolubilization of polymer treatments are well documented (77). Aging and deterioration effects can result in loss of transparency, embrittlement, and cross-linkage formation, thus necessitating the removal of consolidation treatments. Therefore, a cleaning methodology of the degraded and unsuccessful past conservation treatments must be applicable to the whole film.

Femtosecond (*fs*) laser pulses (500 *fs*) at 248 *nm* were generated based on a XeCl excimer pumped dye laser system operating on the principle of a distributed feedback dye laser (DFDL) (76). The maximum energy reached at 248 *nm* in the setup is 4 *mJ* per pulse, while the average pulse-to-pulse fluctuation is 15%. The beam was focused perpendicularly on the sample by means of a quartz lens (f = +300 *mm*) onto an area of approximately 0.006–0.009 cm^2.

Nanosecond (*ns*) laser pulses (30 *ns*) were emitted from a KrF excimer laser system (Lambda-Physik, LPX 200) operating at 248 *nm*. The maximum energy per pulse was 40 *mJ* while the laser beam was focused on the sample by means of a fused silica plano-convex lens (f = +100 *mm* onto a surface of 0.022 cm^2). All irradiation experiments were performed in air.

The irradiated areas were examined for changes and alteration to their morphology under an optical microscope using both transmitted and reflected light. Also, scanning electron microscopy was carried out following metallization with gold (76).

Scanning electron images of an irradiated Elvacite film and Photomicrographs of a vinylic PVA film are shown in Figures 1.10 and 1.11.

In the case of Elvacite and Paraloid B-72, acrylic materials with relatively low glass transition temperatures, the irradiation resulted in craters with blurred edges and pitted crater surfaces. This is associated with the presence of well-defined craters from exploded bubbles in the substrate. The foamy formations are observed even with single pulse irradiation and the size of bubbles ranges from 1 to 20 μm (76).

The vinylic PVA, which has a relatively high glass transition temperature, shows smoother crater surfaces under similar conditions. Bubbling formations are limited, while their sharp and cracked edges are evidence of a more brittle material or more abrupt interaction with the laser irradiation. The diameter of bubble formations was in the range of 10–50 μm (Fig. 5a); in this case, the edges of the crater are relatively clean, while wavy formations on the border of the ablated area imply morphological changes due to diffusion of heat.

in contrast to *ns* irradiation, in the *fs* regime the dominant morphological changes are redeposition and formation of various structures on the surface of the polymeric materials. Bubble formation is limited and was observed mainly on vinylic PVA, which has a high melting point of 200°C and a glass transition temperature of 85°C. Single pulses are sufficient to induce melting with bubble formation. The results are shown in Figure 1.12.

A method of removing contaminates from a painted lacquered or patina surface of an artwork is to use a pulsed laser having a wavelength of between 2.3 and 3.3 μm, or between 5.5 or 6.5 μm, such as an Er:YAG having a wavelength of 2.94 μm, Er:YAG YSG, or a pulsed hydrogen fluoride laser, to stimulate the presence of an

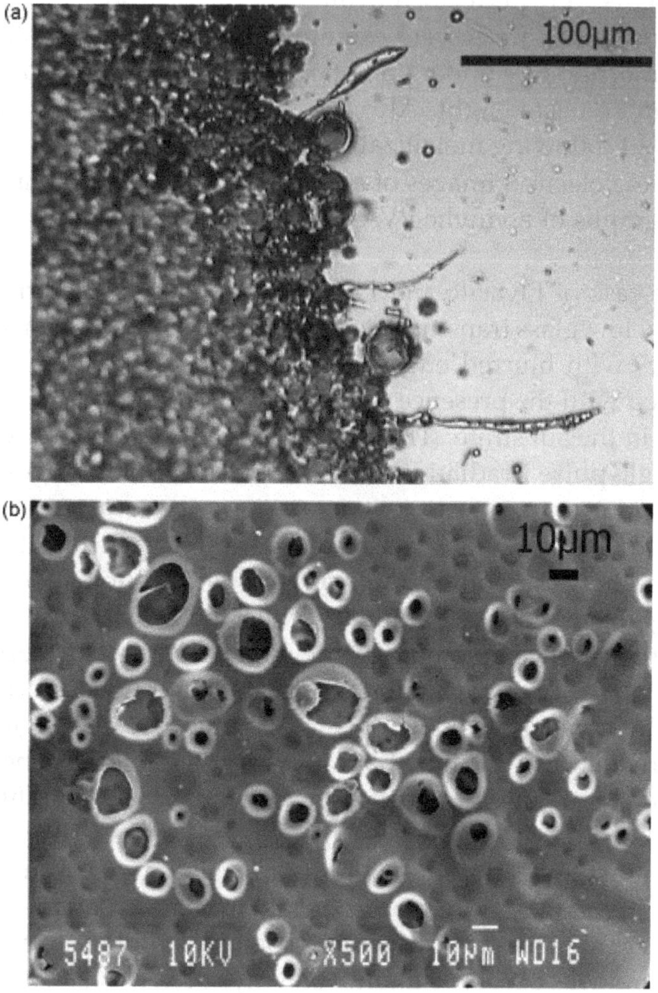

Figure 1.10 Scanning electron images of Elvacite film upon 10 pulses of 248 *nm*, 30 *ns* irradiation (a) on the crater edges (magnification 100×, F = 1730 $mJ\,cm^{-2}$) and (b) on the crater surface (magnification 500×, F = 1270 $mJ\,cm^{-2}$). Reproduced from (76) with permission from Elsevier.

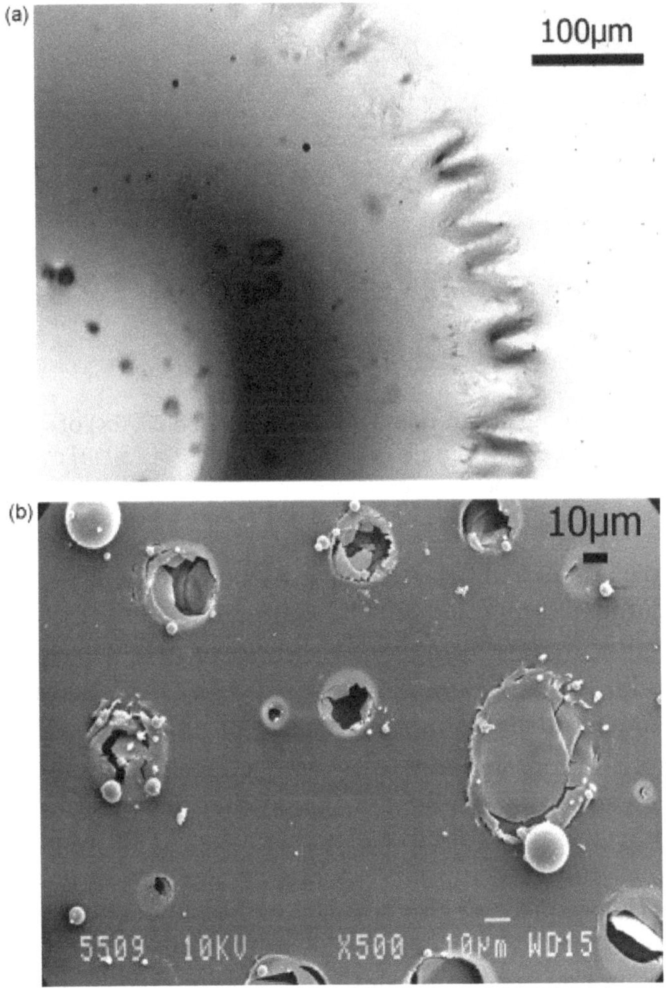

Figure 1.11 Photomicrographs of vinylic PVA film upon 248 *nm*, 30 *ns* irradiation (a) on the crater edges (scanning electron microscope, magnification 500×, single pulse, F = 1530 $mJ\,cm^{-2}$) and (b) on the crater surface (optical microscope, reflectance mode, magnification 250×, 30 pulses, F = 700 $mJ\,cm^{-2}$). Reproduced from (76) with permission from Elsevier.

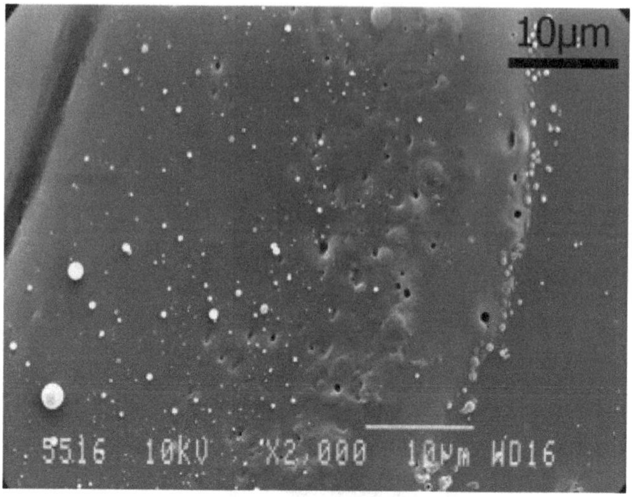

Figure 1.12 Scanning electron image (magnification 2000×) of vinylic PVA film upon 248 *nm*, 500 *fs* single pulse irradiation, F = 150 $mJ\,cm^{-2}$. Reproduced from (76) with permission from Elsevier .

OH bond contained in the contaminant film layer or introduced into the contaminated film with an OH-containing agent applied with a brush, cotton swab or spray (82).

Er:YAG lasers are solid-state lasers whose lasing medium is an erbium-doped yttrium aluminium garnet, i.e., Er:$Y_3Al_5O_{12}$ (83).

Examples of contaminates to be removed include short-chain natural resin varnish, synthetic resins, long-chain polymers, and adhesives, both natural and synthetic, and inorganic salts, both halogens and metals, and all compounds which combine with an OH radical (82). The surface to be removed is covered by a glass cover slip and exposed to the laser energy. The laser energy interacts with the OH group or OH radical present in the contaminant.

The reaction of the laser energy with the OH radical deposits the contaminated film material onto the glass slide cover. Contaminants that remain on the paint surface can then be removed by scalpel or cotton swab using a wetting agent.

This method is a process for improving the cleaning of painted lacquered and patina surfaces by the removal of contaminant films, such as varnish films, polymerized overpaint, and calcium carbonate gesso layers which cover and obscure the painted surface (82).

Gesso is a white paint mixture consisting of a binder mixed with chalk, gypsum, pigment, or any combination of these materials (84,85).

The process of cleaning includes selecting an agent containing an OH bond or radical, such as distilled water, alcohol, including isopropyl alcohol, acetone, and ethanol, for use in cleaning a painted surface. The selected OH-containing agent is applied to a predetermined surface area of an artwork to be cleaned. A selected laser of a predetermined wavelength between 2.3 and 3.3 μm, or 5.5 and 6.5 μm, such as an Erbium:YAG laser having a wavelength of about 2.94 μm, and directing the laser energy onto the area of the painted surface where the OH radical is present at a predetermined energy level for a selected number of laser pulses. Materials loosened by the laser energy interacting with the OH radical is then removed either as dry material or with a wetting agent (82). Several examples have been detailed in the literature (82).

1.1.18 Atomic Oxygen Plasma for Removing Organic Protective Coatings

The combination of an atomic oxygen source for forming an atomic oxygen beam and an equipment for controlling the beam to remove a protective organic coating from the surface of a painting has been described (86).

Here, a painting consisting of inorganic paints and a translucent protective organic layer is subjected to an atomic oxygen beam in a controlled manner, which removes the protective coating without damaging the underlying paint. Previous techniques used to remove organic protective coatings typically consisted of the immersion of the painting in organic solvents or rolling a swab containing the solvent over the surface of the painting.

For example, there is no known solvent which will remove poly(urethene) without damaging the painting. For acrylic lacquer and natural resins, the current cleaning techniques allow contact of either a swab or liquid with the surface of the painting, which alters the surface by removing pigment, smearing the pigment or altering the shape of the canvas surface. Swabbing is unable to remove the lacquer effectively from the low spots on the paintings without damaging the adjoining pigment. Thus, restoration typically invites the damage to the painting (86).

In the process, the painting is placed inside a vacuum chamber and exposed to an atomic oxygen directed beam generated by an electron cyclotron resonance plasma or other atomic oxygen beam source. The atomic oxygen chemically reacts with the organic protective coating, forming a gaseous byproduct which leaves the surface and is pumped out through the vacuum system. The atomic oxygen penetrates the organic protective coating until it comes into contact with the paint pigment particles, which are usually inorganic. The atomic oxygen does not react with the inorganic pigment, and the pigment acts as a shield to the binder layer underneath. The reaction slows as more pigment is exposed because the pigment consists of largely unreactive inorganic material. This continues until all of the organic protective coating is removed from the surface (86).

This method can also be used to remove contaminants from the surface of a painting resulting from, e.g., a fire. These contaminants include carbon soot and other hydrocarbons.

1.1.19 Rigid Gels and Enzyme Cleaning

Hydrolytic enzymes and polysaccharide-based rigid gels are helpful cleaning tools for a variety of polychrome artworks such as moveable paintings, wooden polychrome sculptures, paper and library materials.

Rigid gels have several advantages over conventional fluid gels such as (87):

1. Delivering water or aqueous solutions in a controlled way,
2. Being able to draw into their mesh any particles dissolved upon their application, and
3. Generally not requiring a posttreatment rinsing.

Agarose is the purest galactose-derived gelling material. It produces clear, more transparent gels than the clouded agar gels. This is a particularly important feature when treating delicate supports.

Agaropectin, the other polysaccharide component in agar, is a sulfated polymer. It is a polysaccharide found in agar preparations consisting of *D*-galactose and 3,6-anhydro-*L*-galactose linked by β-1,3 glycosidically. Some of the galactosyl units are sulfated. 3,6-Anhydro-*L*-galactose is shown in Figure 1.13.

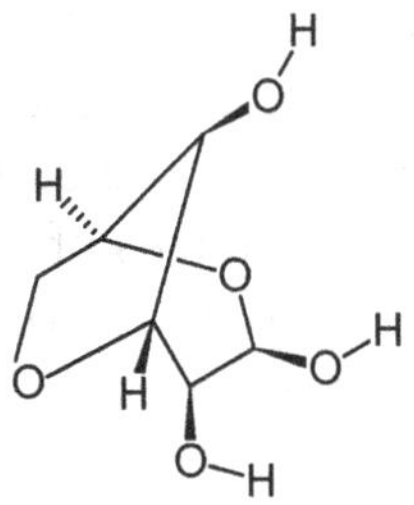

Figure 1.13 3,6-Anhydro-*L*-galactose.

The bulky sulfate groups decrease the pore sizes within the polymer network, thus slowing down the discharge rate of water. Besides applications for the cleaning of surfaces, rigid agar gels in many instances can perform as true humidification systems, capable of the controlled release of water.

According to the requirements of the specific surface treatment, the gel may be left on for a limited time, until the amount of water necessary to obtain the desired results has been released or until the water has been totally discharged. In the latter case, a solid crust remains. This crust is easy to remove because it does not adhere to the surface. The observation of the gradual coloring of the gel as it absorbs the dissolved materials is often used to verify its action.

Hydrolytic enzymes, as they act selectively on specific substrates under mild conditions, can often be a valid alternative to conventional acids and alkalis that is safer to the artwork and less hazardous to the conservator's health.

Proteases, amylases, and esterases hydrolyze peptide bonds in proteins, α-1,4-glucosidic bonds in starch, and ester bonds in esters, respectively. These enzymes have found application in conservation treatments.

For example, the removal of film-forming materials, such as animal and fish glues, gelatins, albumin, casein, starch and starch-containing materials, vegetable glues, drying oils, fats, waxes, and some ester-containing synthetic resins under controlled conditions, is possible (87).

1.1.20 *Cleaning Efficacy of Sponges and Cloths*

The properties and cleaning efficacy of five different sponges and two microfiber moistened cloths were studied on both naturally and artificially aged and soiled oil paint surfaces (88). The cleaning efficacy of these materials were compared to that of conventional cotton swab cleaning with water and saliva. All materials tested showed a good performance in terms of cleaning in comparison to swab rolling with water or saliva. Among all of them, the tested moist isoprene makeup sponge performed best on the artificially aged and soiled surfaces.

1.1.21 *Smart Cleaning by Soft Nanoscience*

Smart cleaning agents are a main issue of modern conservation science. Nanosciences not only provide solutions to this scientific field in terms of new materials but also radically change the approach to problems and challenges. The most innovative nanostructured systems developed in the last decade for the cleaning of artworks have been reviewed (89). Also, case studies have been documented.

Micelles, microemulsions, thickened complex fluids, and responsive gels have been critically analyzed.

The development of these smart nanostructured systems requires the comprehension of their behavior and interactions with other materials down to the nanoscale. Also, the results concerning the mechanism of polymer removal from porous artifacts using nanofluids, such as micelles or microemulsions, have been presented (89).

1.1.22 *Plywood Panels*

In the 1960s, Francesc Artigau painted a series of nine works on plywood panels using a traditional primer of commercial gouache, egg yolk, and colored pencils. After more than 30 years of exposure to rubbish amidst an indeterminate number of dogs, the paintings showed a deplorable state of conservation. The paint and ground layers suffered from the effects of paint loss, damp spots, loss of cohesion of materials, fungi, grime embedded into all levels, erosion, and traces from cleaning tools.

The treatment that was recently done on the paint and ground layers has been reviewed (90).

Agar was chosen as cleaning agent since the components of both the dirt on the paint surface and the paint layer itself were water soluble. The rigid structure of agar allows the water needed in the cleaning process to come into contact with the dirt on the surface while being retained within a gelatinous structure, thus ensuring that the paint layers are not impregnated. It has the additional advantage of not needing to be rinsed afterward as its rigid structure does not leave any residues.

Actually, it was not possible to remove all of the soiling adhered to the paintings by the cleaning process in use because of the degree to which dirt had been absorbed into the strata and the fragility of the ground and paint layers, which did not allow a differentiated treatment of grime. In fact, some of the dirt embedded in the paintings over the years has become part of its composition (90).

1.1.23 Waterborne Emulsion Polymer Paints

In the 1950s, new paints based on waterborne emulsion polymer were used (7). The ability of these paints to be used as incredibly transparent water washes and, at the other extreme, as heavy-bodied, dense gel-like materials, enabled the artists to accomplish things visually that were not possible previously with slower-drying oil-based paint media.

The new media enabled the artists to move aesthetically far from the traditional limitations of oil paints. But on the other hand, the conservation of such paintings became a great challenge. However, some modern paint conservation issues in museums are exacerbated when the same waterborne emulsion polymer paints are used in outdoor public murals (7). Also, the conservation issues in modern and contemporary murals have been reviewed (91).

Studies with ethyl acrylate methacrylate copolymers and butyl acrylate methyl methacrylate copolymers suggested that their lack of hardness at ambient temperatures, along with a tendency to segregate or separate surfactant materials to their surfaces, create a very soft surface which allows easy entrainment of soils. The above monomers are shown in Figure 1.14.

Pure water and nearly all solvent applications to acrylic paint films exacerbate the extraction of paint surfactants, swell or soften the paints. Therefore, severe limits in the surface treatment and

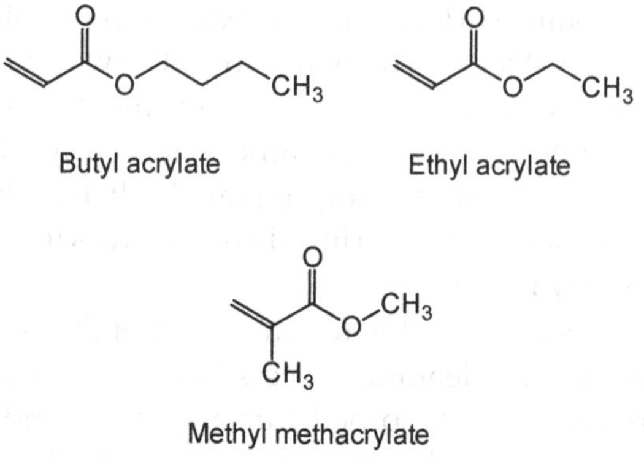

Figure 1.14 Acrylate monomers.

cleaning arise. Solvent and water absorption and swelling of acrylic paint films are due to a number of factors, including the type and amount of materials present within these films and the physical arrangement of the materials within these films.

Absorption and swelling phenomena in these films may be influenced by many factors, such as pigment loading, fillers, age hardening at surfaces, polymer elasticity, film drying rate, film thickness, priming or support layers beneath the acrylic films, and others.

The results of a series of experiments specifically designed to check the pH and conductivity effects of cleaning solutions on acrylic paints have been reported. Golden Artist Colors Heavy Body Tube Colors and ColArt Liquitex Acrylic Tube Colors were used as the model test paints.

Both digital microscopy and liquid chromatography mass spectrometry were used to characterize the physical changes in these paints in terms of surface area increases and to quantify the surfactant moieties extracted.

The results from these extraction experiments suggest that there are clear trends in the swelling of these paint films under short-term immersion conditions as a function of pH and conductivity in the presence of monovalent ions (7).

1.2 Varnishes

Varnishes are the most vulnerable part of old master paintings. Until the mid-19th century these paintings received this final transparent coating. Environmental issues cause traditional picture varnishes to break down rapidly. Autoxidation effects hazes, cracks and yellowing (92).

1.2.1 *Removability of Varnishes*

The removability of naturally aged synthetic picture varnishes has been examined on more than 50 paintings (69). Most of these paintings were coated with either poly(isoamyl methacrylate) or poly(*n*-butyl methacrylate). Since these aged methacrylate coatings were applied up to 50 years before, questions had been posed concerning a possible crosslinking of the polymer chains. This would make them very difficult to remove using solvents or even make them completely insoluble. The respective monomers are shown in Figure 1.15.

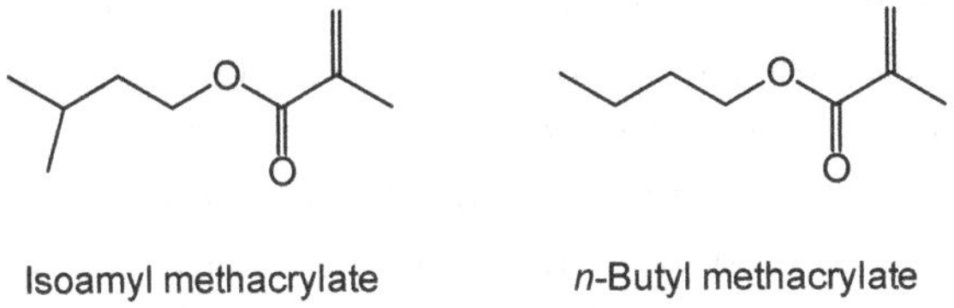

Figure 1.15 Isoamyl methacrylate and *n*-Butyl methacrylate.

These issues were investigated by applying solvents of increasing polarity to the varnish by swab, and analyzing the material removed by infrared (IR) spectroscopy.

It was found that all coatings were removable with solvents, although often fairly polar solvent mixtures were necessary. Only 5 of 25 paintings suspected of being coated with poly(isoamyl methacrylate) were found to be actually coated with this varnish. The varnish was removed with relatively mild solvents, often by swelling.

Further, 31 paintings coated with poly(*n*-butyl methacrylate) were examined (69). Although solvents more polar than acetone were not required for the removal of the coatings, concerns arose about the sensitivity of the paint films of early American paintings to that solvent. In addition, even when these coatings are removable today, the future aging behavior of these varnishes under the conditions of exposure at the National Gallery of Art, where ultraviolet filters are in place, has not been determined. Thus, a continued systematic monitoring of the aging behavior of these varnishes on selected paintings is needed.

1.2.2 *Synthetic Resins for Varnishes*

Synthetic low molecular weight resins have found wide application as varnishes for paintings in the field of art conservation because of their excellent optical characteristics and chemical stability (93). Traditional natural varnish media, such as dammar and mastic, have excellent optical characteristics. However, they are chemically unstable and may become difficult to remove upon aging.

Many synthetic polymeric materials used as varnishes have excellent chemical stability but only poor optical characteristics.

Data on the molecular weight distributions and glass transition temperatures of low molecular weight resins, such as urea aldehyde resins and hydrogenated hydrocarbon resins, have been presented, as well as data on the changes in their molecular weight distributions during exposure as films to a xenon arc lamp for approx. 3,000 *h* (93).

Several industrially manufactured low molecular weight synthetic resins have been investigated for their potential use in picture varnishes (94). The resins were examined by IR spectrometry, gel permeation chromatography, and DSC. Refractive indices were also determined. The stability of films of the resins was tested during accelerated aging in a xenon arc fadeometer and evaluated by means of ultraviolet-visible (UV-vis) and IR spectrometry and solubility measurements. The effects of the hindered amine light stabilizer (HALS) Tinuvin® 292 on the aging of these resins and the ketone resin Laropal® K80, the reduced ketone resin MS2A, and a chemically modified ketone resin, i.e., a ketone-ester, were determined.

Most of these investigated products are considerably more stable than both natural and ketone resins and would, when stabilized with HALS, show considerable longevity even in an environment containing UV light. MS2A and the ketone-ester can also be stabilized effectively by the addition of HALS. Painting conservators have tested the working and aesthetic properties of some of these products (94).

The ketone resin MS2A, the aldehyde resin Laropal® A81, and the hydrocarbon resin Regalrez™ 1094 from Eastman Chemical Resins, Inc., now commonly used in restoration, were characterized by mass spectrometric techniques (95). These resins are commonly used as paint varnishes.

Ketone resins are the condensation products of methyl cyclohexanones and/or cyclohexanones (70). Such condensation can be achieved by use of methanol or formaldehyde in an alkaline medium. Regalrez 1094 is a hydrocarbon resin synthesized from vinyltoluene and α-methylstyrene and subsequent hydrogenation of the unsaturated polymer (96). The compounds are shown in Figure 1.16.

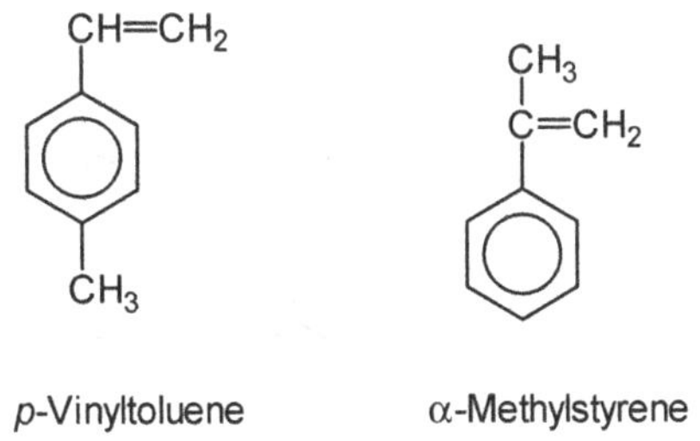

Figure 1.16 Vinyltoluene and α-methylstyrene.

Laropal A81 from BASF is an aldehyde resin synthesized from urea, isobutyraldehyde, and formaldehyde.

Before, the molecular structures of these resins have not been completely elucidated. To improve current knowledge of the chemical composition of these materials, information obtained from GCMS, pyrolysis GC/MS, and electrospray ionization mass spectrometry was combined. These techniques enabled the identification of previously unreported features of the polymer structures (95).

1.2.2.1 *Methylolated Varnishes*

Ketone resins based on condensation reactions involving methyl cyclohexanone and/or cyclohexanone were first patented in the 1920s (97).

MS2A is the chemical reduction product of a complex mixture of methylolation and condensation products formed by reactions involving methyl cyclohexanone, methanol and their derivatives, as shown in Figure 1.17.

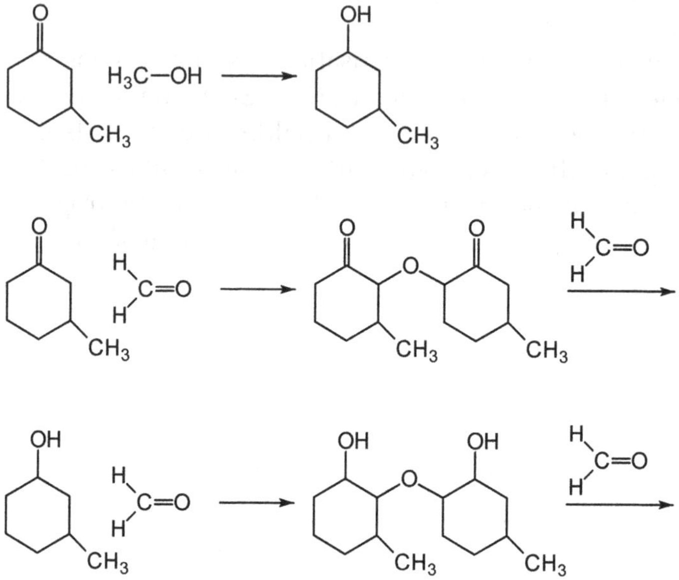

Figure 1.17 Formation of MS2A.

The resins have a high solubility in low polarity solvents coupled with a high refractive index. This enables the formulation of high-solids, low viscosity varnishes which are fairly easy to apply by brush or spray and provide attractive optical properties.

The fast drying rate is an advantage in many cases. On the other hand, the varnish films were found to break down even in museum conditions. A major problem associated with these resins is the tendency of the ketone groups and double bonds to degrade in light, resulting in yellowing and a loss of solubility over time.

The preparation of a MS2A basic varnish for brushing or spraying runs as follows (97):

Preparation 1–1: 50 *g* of an MS2A resin broken down to small granules and suspended in a muslin bag is dissolved in 110 *ml* Stoddards solvent (white spirits to BS 245 having less than 20% aromatics content). The resin dissolves overnight, or more quickly with frequent stirring and a little heating. The final concentration is 31%. 50 *g* of MS2A plus 110 *ml* Stoddard equals a total volume of 160 *ml*

Also, other methods for the preparation of a basic matt varnish solution and a diluted matt varnish as finishing layer have been detailed (97).

1.2.3 *Ionic Liquids for Varnish Removal*

The development of less harmful cleaning methods is a major concern for conservators and restorers. Therefore, the possibility of replacing volatile and toxic organic solvents by ionic liquids could contribute to safer procedures (98).

Ionic liquids are organic salts with melting points below 100°C, constituted by an organic cation and an inorganic or organic anion. Some of the properties of ionic liquids may be useful for the conservation of paintings (99–101).

The low volatility of ionic liquids reduces the risks to health and the environment, their high viscosity may inhibit the penetration into the pictorial layers, and their adjustable miscibility with water or less toxic solvents, such as alcohols, allows their removal with safer solvents (98).

The effects and consequences of possible cleaning operations with ionic liquids have been considered. The effectiveness and risk evaluation was carried out through the application of different types of ionic liquids in mock-ups prepared with various painting materials and varnishes (98).

1.2.4 *Extraction of Soluble Components by a Varnish Solution*

The effects of a varnish applied to an oil paint film have been described. The extraction of soluble material from the paint film by the varnish solution has been investigated (102).

A series of oil paint films were coated with varnishes formulated with a range of resins and solvents. The extraction of fatty acids from the paint film into the varnish layer was measured by GC of the dried varnish film and of samples of the paint film before and after treatment.

The extracted quantities were found to depend not only on the solvent used to apply the varnish but also on the resin component with which the varnishes the were formulated. More polar, oxygenated resins extract more than those made from nonpolar, hydrocarbon resin (102).

1.2.5 *Mastic and Megilp*

Triterpenoid resins, i.e., mastic and dammar, are natural varnishes and are widely used in painting and polychrome sculpture, because they are less prone to yellowing in comparison to other natural resins and also because of their excellent optical properties that are superior to those of more stable synthetic varnishes.

Thus, mastic, which is one of the best natural varnishes, is frequently used as a protective and finishing layer or as a component of oleo and resinous media in paintings. This occurred both in the past and also occurs currently (103).

However, this resin can perform complex deterioration processes, which can change its characteristics and thus the visual aspect of artworks (103).

The alteration processes due to radiation have been widely studied. However, there is a lack of information on the biodeterioration of this natural product. Fungi from collections as well as from oil paintings of the Fine Arts Museum of Granada (Spain) were inoculated onto slides covered with mastic. After an incubation period of 15 *d* the samples were analyzed by GCMS in order to identify the chemical changes. Further, a visual monitoring of the samples was performed to determine the formation of mycelia onto solidified resins. Major changes could be detected in *Chrysonilia sitophila, Phoma herbarum,* and *Penicillium chrysogenum,* showing evidence of alteration processes caused by these microorganisms (103).

The issues of conservation of mastic and megilp varnishes have been exemplified in detail (104).

1.3 Methods and Materials for Conservation

1.3.1 *Microbial Contamination*

One of the biggest problems found in the conservation of a cultural property is to prevent and control the microbial contamination (105,106). In South America, which has a tropical climate with high temperature and relative humidity, it is very common to find artworks with a growth of mold. For a special recently restored Peruvian painting from the 17th century, on which a hard contamination of mold was growing on close to 70% of its area, gamma irradiation appears to be a feasible alternative for biological decontamination and conservation.

However, the radiation can interact with the molecular structure of the material, thus changing its characteristics, especially those related to the color of pigments and the chemical and physical properties of the polymers (105,106).

1.3.1.1 *Ionic Strength*

Actually, a dried acrylic paint film is not completely coalesced. It is an imperfect film that will have physical properties somewhere between those of an impermeable film, a semipermeable membrane, and a spongelike, microporous surface (52).

If an aqueous cleaning system is used on a paint that behaves like an impermeable membrane, it can be reasonably expected that nothing will happen to the paint film when the surface is cleaned.

However, if a cleaning solution is applied to a semipermeable membrane, some interactions between the paint and the liquid will occur by osmotic forces.

If the cleaning solution is applied to a paint that behaves like a microporous surface, the interaction between the solution and components in the paint will be driven by an ordinary diffusion process.

When an aqueous solution is placed on the surface of a microporous paint film, the relative concentration of soluble salts in the paint will try to become in equilibrium with the ionic solutes in the cleaning solution.

If distilled water is placed on the surface of the paint film, the higher ionic strength inside the paint film will cause the diffusion of the ionic species into the water, leaching original material out of the paint film.

When the cleaning solution is of a lower ionic strength than the paint, the cleaning solution is known as hypotonic. If the cleaning solution has more ionic species than the paint film, ions from the cleaning solution will diffuse into the paint film. In this case, the cleaning solution would be hypertonic to the paint film.

If, however, the ionic strengths of both the cleaning solution and the paint film are similar, although there may be some ionic exchange between the solutions, there will be no total movement of ions into or out of the paint layer.

Therefore, according to a porous paint film model, a cleaning solution that is isotonic to the paint is desirable.

When the acrylic paint film is considered as a semipermeable membrane, rather than a microporous film, the observance of osmotic effects can be expected, when a cleaning solution is applied to the paint film.

If distilled water is placed on the acrylic paint surface, the dried acrylic paint film is not completely coalesced. It is an imperfect film that will have physical properties somewhere between those of an impermeable film, a semipermeable membrane, and a spongelike, microporous surface. Such a hypotonic scenario is not desirable.

A difficulty is to determine the ionic concentration of the paint film so that a cleaning solution can be formulated to be isotonic to it.

Measurement of the conductivity and pH of a drop of distilled water placed on the surface for a short period of time, but long enough to allow some sort of equilibrium with the acrylic paint to be reached, gives only imperfect data of the ionic environment of the paint film.

The drop of water is placed first into a single-drop conductivity meter and then into a single-drop pH meter, both of which are commercially available.

Using the complex aqueous cleaning option in the program, the Modular Cleaning Program allows the conservator to control the ionic strength of the test cleaning solutions along with all of the other solution parameters managed in a simple cleaning.

The ionic strength can be lowered by decreasing the amount of buffer and chelating agents added to the test cleaning solution. If a higher ionic strength is required, an ionic buffer, typically dilute NaCl, can be added to test the cleaning solution (52).

The results of testing the cleaning efficiency through high-throughput testing indicated that for acrylic paints the optimum cleaning efficiency is achieved at a concentration of 0.7% triammonium citrate, or 29 mM. With this information, the conservator can specify that chelating agents should be used at a 29 mM concentration when used for acrylic paints (52,107).

1.3.2 *Oil Paintings*

Oil paintings on canvas have only a limited time period before they start to disintegrate (108). Although the paint film may endure for centuries, the canvas undergoes a continuous and fairly rapid chemical and physical decomposition. Consequently, restoration is often required within decades.

Conventionally, painting restorers laminate a new canvas backing over the deteriorated canvas in a process called lining. After a period of time, the second canvas deteriorates and must be replaced, i.e., relining. This repeated process dictates the use of a reversible adhesive, since the previously laminated canvas backing must be removed before applying a new canvas backing. Restorers traditionally use hot melt adhesives for the lining and relining processes, for these adhesives lend themselves to easy removal (108). Adhesives and consolidants in painting conservation have been detailed in a monograph (109).

Historically, beeswax has been the hot melt adhesive used, but in recent years some restorers have tried synthetic hot melt adhesives. Because it is recognized that heating can damage the paint film, some restoration has been performed using pressure-sensitive adhesives at normal room temperatures (108).

Instead of canvas, it has been proposed to use a flexible substrate layer of fine mesh synthetic material such as a woven polyester monofilament, fiberglass, or a polyester mylar (110). The substrate may be fixed to the rear face of the original canvas of an oil painting by a silicone pressure-sensitive transfer adhesive applied at room temperature.

Alternatively, a heat activated adhesive, such as ethylene-vinyl acetate or a wax resin, has been suggested. The substrate can be used alone, or it can be backed with a canvas or poplin lining.

It is well recognized that any work done on a painting damages it. The lining and relining processes expose the painting to physical and chemical damages (108). Hot melt adhesives thermally damage the paint film, both when applied and when subsequently removed. Adhesives that require solvents for removal inflict chemical damage to the paint film. Water-based adhesives cause the canvas to contract and expand, thus stressing the paint film.

The deterioration of the original canvas is the most damaging of all to the paint film. As the canvas decomposes, it loses its original function, that of supporting the paint film. This causes the film to crack and ultimately to separate from the canvas. Contractions and dimensional changes in the original canvas introduce stresses in the paint film that further weaken it.

The application of a second canvas backing and filling of voids with adhesives or waxes does not alleviate the stresses induced by the dimensional changes in the original canvas. The added lining simply provides a temporary treatment to the painting. As long as the original canvas is intact, it will continue to deteriorate and the paint film will continue to lose its support. The deteriorated canvas will continue to stress the paint film. Over time, many works of art are lost long before the paint film deteriorates, due to the instability of the canvas. Even the use of synthetic materials for lining and relining does not appreciably help the situation, since the deterioration of the original canvas continues to introduce stresses in the paint film (108).

A process has been proposed that makes it possible to produce an integrated structure to restore and permanently preserve an aged oil painting on canvas (108). The process involves first eliminating the stresses in the paint film caused by the deterioration of the old canvas and assuring that no additional stresses will be subsequently introduced by further deterioration of the old canvas. This objective is accomplished by abrading or otherwise breaking down the canvas weave. The paint film, ground layer, and old canvas remnants are then integrated with a new non-canvas support into a single composite structure by means of a coherant. A coherant denotes a material capable of penetrating through the remnants of old canvas, filling the voids on the back of the paint film, and forming a continuous bond with the paint film, likewise capable of filling any voids in the new support and forming a continuous bond with the new support, and not significantly oxidizing or otherwise deteriorating during the expected lifetime of the paint film when exposed to normal atmospheric conditions.

In particular, the process comprises (108):

1. Removing at least a portion of the back surface of the canvas layer sufficient enough to break the weave so as to eliminate any stress introduced by the canvas layer in the paint layer,
2. Providing a dimensionally stable new backing of a material having a life expectancy comparable to the expected life of the paint in the paint layer, the backing having a front face and a back face,
3. Applying a uniform coating of a liquid coherant to at least one of the backs of the painting and the front face of the new backing, the liquid coherant being capable of penetrating the remaining canvas layer and ground layer and of curing at ambient temperature to a form having a life expectancy comparable to that of the new backing material and of securely bonding, in the cured form, to the paint layer and to the new backing material without any deleterious effect on the paint layer, and
4. Compressing the painting and the new backing together by applying uniform pressure to the face surface of the paint layer and to the rear face of the backing until the coherant has cured.

1.3.3 *Organic Materials*

1.3.3.1 *Natural Materials*

Among the naturally occurring products, proteins, lipids, polysaccharides and terpenoids are widespread in art objects. Protein materials, such as animal and fish glue, eggs and casein, are incorporated into paintings as binding media (111).

Glues can be obtained by dissolving collagen, the fibrous protein of connective tissues in animals and fish. Egg white and egg yolk have a quite complex composition in terms of their individual proteins, including glycoproteins, ovoalbumin and lysozyme in the white and phosphorous-containing lipids, such as phosphovitin, and also lipovitellins, lipoproteins and livetins in egg yolk.

Casein is a phosphoprotein complex that is precipitated by the acidification of skimmed milk that has been mainly applied as glue and eventually as a tempera medium.

The drying oils included in the glyceride group of lipids, as they are mixtures of mixed triglycerides, are of basic importance to oil painting and the coating technology. The various materials known as waxes are, in practice, a heterogeneous group containing mixtures of hydrocarbons, acids, alcohols, esters, sterols and terpenoids (111).

They have been used as adhesives and surface coatings, and particularly as waterproofing agents and, occasionally, as ingredients of binding media. Polysaccharide gums exuded by trees when wounded, have preferably been used as a medium for water colors, miniatures and manuscript illumination as well as adhesives. Although the chemistry of natural resins is diverse, most are mainly composed of terpenoid compounds, which are formally considered compounds made up of units of isoprene.

Essential oils are mainly composed of monoterpenoids and sesquiterpenoids and have been used as plasticizers of varnishes. Natural resins contain diterpenoids or triterpenoids as well as monoterpenoids and sesquiterpenoids, to which they owe their initial fluidity. They have been widely used as varnishes and, to a lesser extent, as binding media (111).

According to Mills and White (112), bituminous materials are usually classed among natural products such as bitumen, pyrobitumens and asphalt, and as artificial products made by pyrolysis of wood, coal or resin, such as tars and pitches.

The chemistry of bituminous materials is complex and includes long chain hydrocarbons, terpenoids or aromatic compounds such as phenols. These soft products have been used as adhesives, mortars and coatings.

Besides all the above-mentioned materials, a number of compounds have also been used by artisans as both coatings, i.e., lacquers, and additives of binding media such as emulsifiers (oxgall) and, some more occasionally, as plasticizers (honey) or adhesives (garlic) (111).

1.3.3.2 Synthetic Materials

In addition to natural products, synthetic polymers were introduced into art and art conservation a few years after the invention of the first synthetic polymer, cellulose nitrate, in 1846 by Schönbein.

At that time, Alexander Parkes displayed a variety of molded cellulose nitrate objects at the International Exhibition in London. In the 20th century, the use of synthetic polymers has become widespread, and nowadays they are found not only as materials forming art objects but also as adhesives, consolidants, inpainting varnishes or fillers of missing parts used in restoration works (113).

Among the wide range of synthetic resins with applications in art and conservation, acrylic, PVA and poly(cyclohexanone) are extensively used as artists' painting media or picture varnishes (111)
.

Additionally, it has become a common practice for artists to adapt materials manufactured for other commercial areas such as the decorative or industrial field. This is the case of alkyd resins developed for household and industrial paints and coatings or poly(urethane)s, epoxies, silicone resins and nitrocellulose.

These products have been chosen because they offer a number of advantages over traditional materials when used as binding media and varnishes, such as long-term mechanical and optical stability, and shorter drying times. Some of them offer the feasibility of being used in emulsion, thus avoiding toxic ingredients such as organic solvents. However, the long-term aging of many of them has not yet been completely evaluated (111).

1.3.3.3 Degradation Mechanisms

The stability of an organic compound firstly depends on its structure or on the strength of its different chemical bonds, which determines the alteration mechanisms observed in organic materials (111) .

Deterioration of artists' media and varnishes is mainly a result of oxidation processes involving free radicals and hydrolytic processes of an ionic type. Autoxidation reactions are initiated by thermal or photochemical energy input and more rarely by ionizing radiation without oxygen intervening. An empirical measure of the extent of a photochemical reaction taking place on a varnish or medium is provided by the quantum yield of the photochemical process.

This can be expressed as the ratio of the number of macromolecules undergoing a chain scission divided by the number of quanta absorbed by the organic materials.

Also, the diffusion of oxygen into the varnish or paint film is an important factor which determines the deterioration process. The starvation depth, defined as the depth into the film at which the oxygen concentration was reduced to 10% of that required to maintain the reaction rate, is another parameter used to assess the extent and mechanisms of an oxidative process.

Apart from thermal or photochemical sources, other agents of deterioration have been assessed. Metal oxides and other salts present in pigments can have a notable effect on varnishes and media as they promote light-induced oxidation reactions. In contrast, other pigments such as carbon black have a protecting influence as they prevent the penetration of light. Certain compounds that are present in the varnish or medium in small amounts can act as sensitizers by absorbing the light and by transferring this energy to substrate molecules. The ozone is found among oxidizing atmospheric agents (111).

Hydrolysis reactions are usually catalyzed by acids and also other catalysts, which can be present in the atmosphere as pollutants. This process is particularly important for proteins, drying oils and polysaccharides as it results in a significant change of the chemical and mechanical properties of the films.

Due to the biodegradative activity of molds, bacteria and other higher organisms, biochemical or enzymatic processes take place on the artwork, which are another frequent form of deterioration.

Deterioration of varnishes and binding media results in chemical changes such as crosslinking reactions between the polymeric chains, chain scissioning, oxidation of the main chains or side groups and the breakdown of molecules. This is often accompanied by the formation of highly oxidized products. Such structural changes result in an increase in the insolubility and polarity of the material, a reduction of the strength, and a change in color (111).

1.3.3.4 *Microchemical Tests*

The first chemical investigations into the organic components of historical materials were done in the late 18th century and were simple solubility tests consisting of wiping the surface of the object with solvents in an attempt to characterize their composition according to their solubility properties (114).

Other tests for determining some characteristic physical properties of the medium, such as the melting point (115) and also the microchemical test, were applied in that early period.

Solubility tests can be helpful in the characterization of media, but only to such an extent as the solubility parameters of the materials being tested remain fairly well defined and differ from one another. Nevertheless, the organic materials present in artworks degrade easily so that the overlapping of exhibited solubility parameters is briefly achieved, thus obstructing the identification of binding media.

Most of the microchemical tests used for the analysis of paintings originate from the general histology practice. This technique takes advantage of the affinity of specific functional groups for certain highly colored biological stains, thus allowing for the differentiation of organic bindings.

Apart from their simplicity, another advantage of microchemical tests is the feasibility of identifying the paint media within the structure of the pictures as these tests can be carried out on cross sections, and the reaction can be observed with the help of a stereo microscope. However, this technique is limited by its low specificity since only a poor discrimination can be obtained between the different materials within the same group (111). Such materials are albuminoids, casein or gelatine within the proteinaceous materials.

Already in 1905, microchemical tests based on staining cross sections with methyl violet and iodeosine were reported (116). These tests were used to identify oleaginous and proteinaceous media.

1.3.4 *Poly(vinyl acetate) Paints*

The photochemistry of a PVAc homopolymer, of PVAc mixtures with selected pigments and fillers, and of accurate historic reproductions based on the colors prepared by the Portuguese artist Joaquim Rodrigo and on those supplied by the company A Favrel Lisbonense was studied (117).

The systems, applied as films, were irradiated at a wavelength of greater than 300 *nm* and the changes were assessed by size exclusion chromatography, infrared spectroscopy, and colorimetry.

PVAc as homopolymer or as an emulsion paint proved to be very stable to light, and after 5000 *h* of irradiation no gel fraction was observed. Φ_R (λ_{irr} = 313 *nm*) for chain scission for PVAc was determined to be 7×10^{-8}. This value indicates that the mechanisms operating when irradiating at λ = 300 *nm* are different from those with an irradiation involving 254 *nm*. Also, no side group scission was observed, and main chain scission is the foremost photodegradation mechanism. Further, the metal ions present in the pigments do not affect the photochemical stability of the polymer (117).

1.3.5 *Pressure-Sensitive Adhesives*

In the conservation of paintings on canvas, the original support can be consolidated by attaching a reinforcing fabric to the back of the painting (118). Some of the risks inherent in traditional lining techniques can be eliminated by the use of a pressure-sensitive adhesive.

Adhesives based on acrylic esters were selected because of their long-term stability. The adhesives were subjected to thermo-oxidative aging and changes in their chemical properties and their surface and adhesive properties were studied using FTIR spectrophotometry, goniometry, and peel tests at 180°.

The results indicated that thermo-oxidative aging causes three types of alterations in the adhesives (118):

1. Dehydration,
2. Oxidation, and
3. Crosslinking.

Although the extent of these phenomena is small, the loss of adhesive properties is significant, due to changes in the viscoelastic properties of the glues. However, some of the adhesives retain properties which satisfy the criteria required in conservation (118).

1.3.6 *Microcrystalline Cellulose Composites*

Thin adhesive composite films were prepared by melt-compounding and compression molding of a thermoplastic resin (Paraloid B-72), which is widely used for art protection and restoration with a microcrystalline cellulose powder (MCC) (119).

To simulate an oil painting restoration work, MCC-based composites were applied as lining adhesives on two kinds of canvases (English linen and woven polyester). Interestingly, single-lap shear tests both in quasi-static and creep conditions showed a remarkable stabilizing effect provided by MCC addition. Post-mortem microstructural analysis of the overlap adhesive area proved how MCC introduction did not change the fracture behavior of the bonded joints.

The visual assessment of transparency and color measurements evidenced a chromatic variation of MCC-based films with a filler amount of 30%, while, UV-vis analysis showed a decrease of relative transmittance in proportion to the MCC content. Additionally, rheological tests highlighted a viscosity increase for Paraloid B-72 in both dry and wet conditions at 23°C and a relative humidity of 55%) as the MCC content increases (119).

1.3.7 *Nanoscience for Art Conservation*

Nanoscience for the conservation of works of art has become an important issue (120). Techniques based on nanotechnology are replacing traditional treatments. The scientific and technical background to understand, to produce, and to apply nanosystems for conservation to works of art has been detailed. Numerous case studies have been presented show how to handle these systems.

The interest in microemulsions arises from the possibility of controlling through confinement the activity of hydrophilic/hydrophobic molecules in a nonpolar or polar matrix (121). The microstructural properties of microemulsions have been characterized extensively. However, investigations for the application of microemulsions have mainly been focused on their use as nanoreactors in catalysis and biocatalysis, and their use in drug delivery.

The usage of microemulsions for the conservation of cultural heritage has been investigated (121). Systems have been designed with low amounts of volatile oils as low-impact cleaning tools for the removal of aged polymeric coatings from the surface of paintings.

In this field, besides the ecological issues, also of interest is the potential of oil-in-water microemulsions embedded into an aqueous polymeric dispersion, or bicontinuous microemulsions in which the oil phase is gelled. A *p*-xylene-in-water microemulsion embedded in a hydrophobically modified 2-hydroxyethyl cellulose has been designed.

Such a system has advantages in the cleaning of a work of art with respect to traditional methods (121):

- The penetration of the cleaning agent into the porous matrix constituting the bulk of the painted layer is minimized as a result of the very high viscosity.
- There is a drastic decrease in environmental impact, since the microemulsion is mainly composed of water (more than 85%), and the evaporation of the volatile active oil-dispersed phase (*p*-xylene) is reduced by the high retention of the microemulsion.
- Residues can be removed readily from the paint surface.
- The systems are optically transparent, and the cleaning action can be monitored visually and controlled during the application.

1.3.7.1 Nanolimes

Nanolime dispersed in 2-propanol has been extensively used for the consolidation of wall paintings.

The knowledge of the advances of this methodology dealing with all the possible effects associated with the nanolime new material in conservation is fundamental to assess and improve the technique (122).

Four different dispersions of $Ca(OH)_2$ nanoparticles were characterized by small angle X-rays scattering technique and transmission electron microscopy in order to get information on size, shape, polydispersity, agglomeration, and crystal structure of the particles (122). After analysis the dispersions were tested in two different case studies, the Carceri dello Steri in Palermo with their graffiti and the 18th century lunettes at the SS. Giuda e Simone Cloister, Corniola (Empoli) with their lime-based mural paintings.

The treated samples were characterized with regard to either their morphology and surface chemical composition with an SEM coupled with energy dispersive X-ray spectroscopy, or their water absorption and mechanical properties. The results showed that all the dispersions were nanometrically structured and their application succeeded in recovering the mechanical properties of the painting or graffito layers, and not altering their permeability to water and keeping the wall transpiration (122).

1.3.7.2 *Microemulsions*

A physicochemical investigation of oil-in-water microemulsions and micellar solutions formulated to solubilize acrylic and vinyl polymers has been presented (73).

Three different four-component systems have been tested. The systems are detailed in Table 1.6.

Table 1.6 Four-component systems to solubilize acrylic and vinyl polymers (73).

System 1	System 2	System 3
Water	Water	Water
p-Xylene	Propylene carbonate	*p*-Xylene
Sodium dodecylsulfate	Sodium dodecylsulfate	Tween 20
1-Pentanol	1-Pentanol	1,2-Ethanediol

The application of these microemulsion or micellar systems as solubilizing agents for acrylic and vinyl polymers from works of art (mainly wall paintings) and monuments (stones) has been investigated. Acrylic and vinyl polymers have been extensively used in the past decades, and are still used, for art conservation. The aging produced both yellowing and serious degradation of the painted layers or of the stone surfaces, imposing their removal from the artistic and architectonic surfaces. Contact angle measurements, FTIR, and SEM/energy dispersive X-ray spectroscopy results showed that the microemulsions and micellar solutions investigated are very effective in removing acrylic or vinyl polymeric resins from several solid surfaces.

Hydrophobic acrylic copolymers could be completely removed from a Renaissance fresco by Spinello Aretino (used during a restoration performed in the sixties) in the Cappella Guasconi in San Francesco Cathedral, Arezzo, Italy, and PVAc resins (used in a restoration performed during the fifties) from Renaissance frescoes decorating the external walls of the Cathedral of Conegliano, Italy.

The experiments showed that the nanocontainers route represents a safe and very efficient method for removing an aged polymer from surfaces of works of art (73).

The elimination of hydrophobic materials from the surface of wall paintings can be done as follows (74). Nanostructured systems with well-characterized physicochemical properties are an alternative to unsupported organic solvents for solubilizing substances on porous painted surfaces.

Due to the wide range of components that can be used in their preparation, these systems can potentially be used to remove a wide range of different types of organic materials.

Oil-in-water microemulsions and micellar solutions as cleaning agents for wall paintings have been tested (74). The efficacy of these systems has been tested in several Italian conservation workshops for the removal of various hydrophobic materials. In particular, the compositions have been successfully used during the conservation of the frescoes in the Loggia del Bigallo in Florence, and for the removal of various synthetic organic resins from the surface of the Filippo Lippi frescoes in the apse of the Cappella Maggiore in the Prato Cathedral.

In addition, oil-in-water microemulsions have also been shown to be effective cleaning agents for the removal of oil-based and acrylic-based paint used by vandals to deface an 18th century a secco painting in the Villa del Barone in Prato. The results have been detailed by photographic documentation and FTIR spectra (74).

In the past, restorations performed with acrylic and vinyl polymers showed detrimental effects to wall paintings that lead to the complete disfiguration of the painted surfaces (123). The removal of these materials performed with the traditional solvent-based methodology represents a challenge to the conservators and usually achieves very poor results.

A new palette, nowadays available to restorers, has been reviewed (123). This is based on microemulsions, micellar systems, and physical and chemical gels specifically formulated for the cleaning of cultural heritage artefacts.

The restoration of the Renaissance paintings by Masaccio, Masolino, and Lippi in the Brancacci Chapel in Florence (1984–1990) represents the first case study where microemulsions were used for conservation purposes (124, 125).

In 2007 some wall painting decorations in the Oratory of San Nicola al Ceppo, devastated by the 1966 flood of the Arno River in Florence, were restored by removing the patinas left after the flood (126). The painted surfaces were covered by a crust of gypsum efflorescences mixed with the residuals of oil fuel from the flooding water. Sixty years after the event, it was observed that the aging of this blend of hydrocarbons and other materials resulted in an extensive crosslinking of the organic materials and also an insolubilization of the hydrophobic layer to most organic solvents. So, it was necessary to combine the action of a solvent capable of swelling and partially solubilizing the coating and of chemically attacking the gypsum patina. This twofold purpose was successfully achieved by the use of an oil-in-water microemulsion based on 1% xylene for the swelling of the organic components which was dispersed in an ammonium carbonate solution for the solubilization of gypsum. The composition was used as a continuous phase (123).

Microemulsions specifically tailored to remove acrylic polymers have also been used to remove unwanted graffiti, i.e., paintings, writings, or drawings from walls, stones and monuments made by using common spray varnishes (74).

A physicochemical investigation of oil-in-water microemulsions and micellar solutions formulated to solubilize acrylic and vinyl polymers has been done (73). Some four-component systems are shown in Table 1.7.

Table 1.7 Oil-in-water microemulsions (73).

Compounds			
Water	*p*-Xylene	Sodium dodecylsulfate	1-Pentanol
Water	Propylene carbonate	Sodium dodecylsulfate	1-Pentanol
Water	*p*-Xylene	Tween 20	1,2-Ethanediol

Tween 20® is poly(oxyethylene) (20) sorbitan monolaurate and is shown in Figure 1.18.

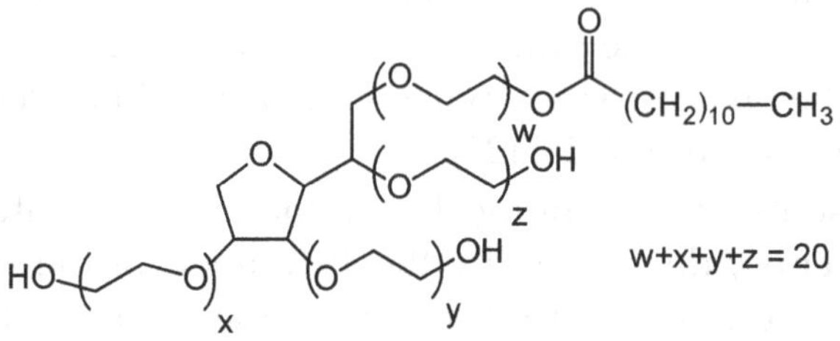

Figure 1.18 Tween 20.

The application of these microemulsion systems or micellar systems as solubilizing agents for acrylic and vinyl polymers for wall paintings and stones has been investigated. Actually, acrylic and vinyl polymers have been extensively used in past decades for art conservation.

The aging produced both yellowing and serious degradation of the painted layers or of the stone surfaces, imposing their removal from the artistic and architectonic surfaces. Contact angle measurements, FTIR, and SEM results showed that the microemulsions and micellar solutions investigated were very effective in removing acrylic or vinyl polymeric resins from several solid surfaces.

Hydrophobic acrylic copolymers could be completely removed from a Renaissance fresco by Spinello Aretino in the Cappella Guasconi in San Francesco Cathedral, Arezzo, Italy, and PVAc resins from Renaissance frescoes decorating the external walls of the Cathedral of Conegliano, Northeast Italy.

Thus, the nanocontainers route represents a new, safe, and very efficient method for removing aged polymer from surfaces of works of art, otherwise condemned to complete loss (73).

1.3.8 Consolidating Wall Paintings Based on Dispersions of Lime in Alcohol

Dispersions of calcium hydroxide in short-chain aliphatic alcohols were studied in order to test a pre-consolidation and consolidation methodology for wall paintings (127). The choice of lime, i.e., a paste consisting of calcium hydroxide plus water as the consolidant, was suggested by physicochemical compatibility criteria, as lime was the original binder in the wall paintings treated.

Many problems have been encountered when calcium hydroxide was used for consolidating wall paintings. Saturated solutions are too weak to act as consolidant, due to the low solubility of calcium hydroxide in water. In addition, suspensions of calcium hydroxide in water are too unstable to be applied to unprotected painting surfaces.

However, the use of calcium hydroxide dispersions in 1-propanol has been proposed (127). These alcohol dispersions were found to be much more stable than suspensions in water.

Submicrometer $Ca(OH)_2$ crystalline particles have been obtained by mixing equal volumes of NaOH and $CaCl_2$ aqueous solutions with different degrees of supersaturation, from 2 to 10, and in the temperature range of 60°C to 90°C (128,129). The $Ca(OH)_2$ particles are crystalline and have the appearance of hexagonal platelets with thicknesses of only a few nanometers.

It was found that 1,2-ethanediol and 1,2-propanediol remained adsorbed onto the nanoparticles, which caused an aggregation, thus forming micron-sized agglomerates. 1,2-Ethanediol and 1,2-propanediol are shown in Figure 1.19.

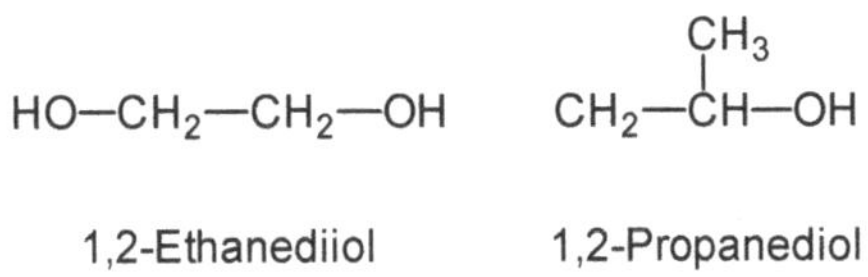

Figure 1.19 Diol compounds.

Their removal, with subsequent dispersion of the nanosized units, was achieved by peptization with 2-propanol in an ultrasonic bath (130).

Although it was believed that $Ca(OH)_2$ particles were inert when dispersed in short-chain alcohols, it has been recently shown that they can undergo transformation into calcium alkoxides (131). The mechanism and kinetics of such a phase transformation as well as its effect on the performance of nanolimes have been elucidated. The $Ca(OH)_2$ particles formed after lime slaking react with ethanol and isopropanol and partially transform into calcium ethoxide and calcium isopropoxide, respectively. These compounds are shown in Figure 1.20.

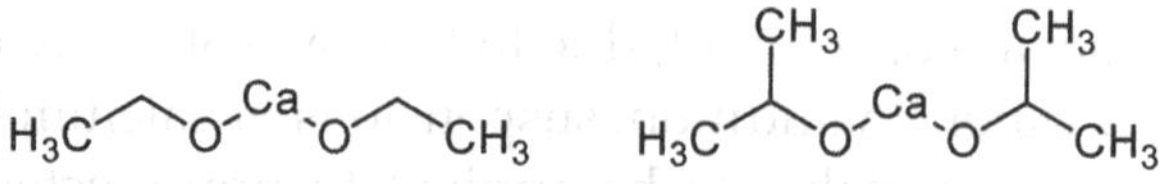

Figure 1.20 Calcium ethoxide and calcium isopropoxide.

The consolidating power of alcohol dispersions on specimens that simulate different degrees of decohesion was examined by X-ray diffractometry and FTIR spectroscopy. Their surface properties were characterized by studying capillary rise, water permeability and mercury porosity. Also, the mechanical properties of the specimens were determined using sclerometric resistance and the Scotch Tape test (132, 133).

The results indicated that the lime/alcohol dispersions produced an excellent consolidation (127).

1.3.9 *Hindered Amine Light Stabilizers*

Oligomeric or polymeric HALS compounds can be used to modify the rheology of materials used for art conservation and restoration (134). Here a lyotropic liquid crystal phase is formed, the rheology of which can be modified by varying the amount of the HALS as well as the solvent or the resin itself.

Hindered amine compounds in oligomeric or polymeric form can, with suitable solvents, be formed into liquid crystal phase gels which have remarkable properties as rheology modifiers for organic solvents and polymer solutions compositions, and in particular retouching or inpainting media and paints used in the conservation of fine art (134).

The lyotropic liquid crystals of the gel consist of a hindered amine compound, solvent molecules, and may contain a polymer depending on the nature of the resin used. Lyotropic liquid crystal molecules are amphiphilic compounds characterized by elongated molecules with hydrophilic and lipophilic ends, the molecules clustering in various configurations in solution, according to concentration and temperature, and depending on the type of solvent used.

At a low concentration, solutions of lyotropic liquid crystal molecules are unremarkable, with solute molecules distributed randomly throughout the solvent. At higher concentrations the molecules begin to arrange themselves into clusters called micelles, which may take the shape of hollow spheres, rods or disks according to their concentration.

At moderate concentrations in a polar solvent such as water, the molecules tend to cluster in hollow spheres, the surfaces of which are formed by a layer of polar heads dissolved in the water with the inner portion comprising hydrophobic tails screened from the water by the hydrophilic heads. The micelles in turn tend to arrange themselves in regular arrays similar to a cubic crystal lattice.

A similar phenomena occur also in nonpolar solvents, but the orientation of the molecules is reversed. In the case of the oligomeric or polymeric molecules the micelles are anisometric and, depending on the concentration, conditions, type of solvent and additional solutes, arrange themselves, giving a variety of structures.

As the concentration increases, a discontinuous cubic phase (the micellar phase) gives way to a hexagonal columnar phase (the middle phase), which in turn gives way to a bicontinuous cubic phase and then a lamellar phase or bilayer, with a double layer of molecules aligned in a polar solvent with the polar heads of each layer facing away from one another and the nonpolar tails facing one another, the orientation again being reversed in a nonpolar solvent.

Because the sheet-like layers can slide easily past each other, this phase is less viscous than the hexagonal phase, at least in the direction of sliding, despite the lower solvent content.

The HALS compounds contain a piperidine ring with alkyl substituents on either side of the ring nitrogen.

One example of such hindered amine is poly(*N*-β-hydroxyethyl-2,2,6,6-tetramethyl4-hydroxy-piperidyl succinate), or a condensation polymer of butanedioic acid with 4-hydroxy-2,2,6,6-tetramethyl-1-piperidineethanol, cf. Figure 1.21.

Preferred solvents are propylene glycol monomethyl ether (1-methoxy-2-propanol), propylene glycol *n*-propyl ether (1-propoxy-2-propanol), and dipropyleneglycol methyl ether (1-(2-methoxyiso-

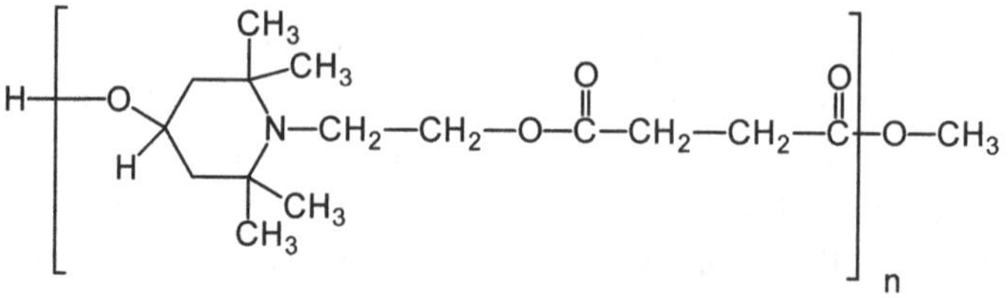

Figure 1.21 Condensation polymer of butanedioic acid with 4-hydroxy-2,2,6,6-tetramethyl-1-piperidineethanol.

propoxy)-2-propanol) (134). These solvents are shown in Figure 1.22.

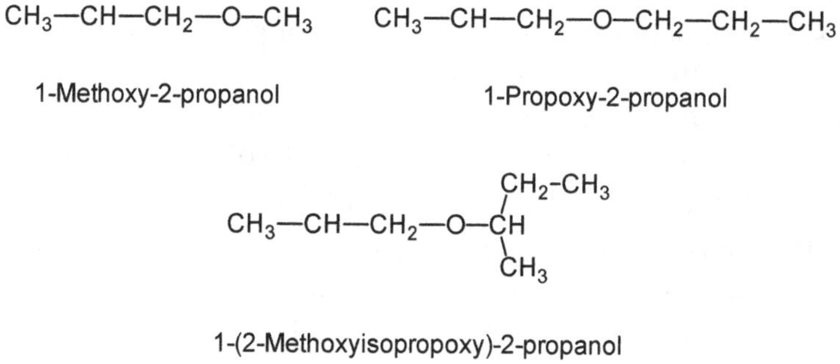

Figure 1.22 Propanol solvents.

1.3.10 Enzymes

Lipases, hydrolytic enzymes that act on glycerol-ester bonds, are often used in conservation for their ability to degrade aged oil films, as a nontoxic and often less aggressive alternative to highly polar organic solvents or alkaline mixtures (75). One such enzyme has been used to remove layers of an aged acrylic resin, i.e., Paraloid B-72, in two instances, a 15th century tempera painting on panel and a 19th century oil painting on canvas.

For a lipase treatment, stone samples have been coated by immersion in a solution containing the lipase enzyme (0.5 $mg\,ml^{-1}$) for a contact time ranging from 2 to 7 *min*, at 38°C (135). The poultice solution is prepared by adding lipase to a carbonate buffer solution of 0.1 *M* at a pH of 8.7 also containing ammonium carbonate at a concentration of 0.1 *M*, where these buffered media represented the best conditions for the lipase enzymatic activity. After cleaning, the stone samples are sequentially washed with distilled water and carbonate buffer solution to remove the excess of the applied lipase enzyme.

1.3.11 *γ-Radiation and Polymers*

A Peruvian painting from the 17th century, which was restored and then contaminated by mold has been studied (136). The painting received various unsuccessful treatments. Therefore, radiation process has been suggested as an alternative. This is an effective technology for decontamination and conservation purposes.

The influence of the irradiation process on the original painting and on the products used in the restoration process has been studied. These products were irradiated with ^{60}Co gamma rays applying doses in the range of 6-25 *kGy*.

Before and after irradiation, the polymeric materials were characterized by thermal analysis techniques. The color of the pigments irradiated and non-irradiated were compared by spectrophotometric analysis. Small samples removed from the original painting were also irradiated and investigated. The results obtained allowed the conclusion that the irradiation with the appropriated dose of 6.0 *kGy*, according to the literature, would not damage the restored painting.

1.3.12 *Partially Hydrolyzed Poly(vinyl acetate) and Borax Gels*

Gels based on PVA have been investigated in order to expand the range of options for conservators. PVA can be crosslinked with borax to form well-characterized hydrogels (137). However, here the amount of gelation is comparatively small.

On the other hand, PVAc with reduced hydrolysis fractions can form borax crosslinked gels with a large mole fraction of methanol, ethanol, 1-propanol, 2-propanol, acetone, or *N*-methyl pyrrolidinone, cf. Figure 1.23, as a cosolvent with water.

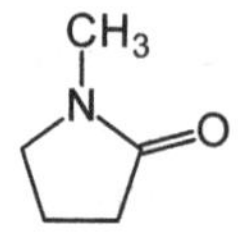

Figure 1.23 *N*-Methyl pyrrolidinone.

These gels can be made nonsticky, transparent, pliant, and stable over periods of at least weeks in closed vessels at room temperature. So, they have a great potential for the removal of dirt and other coatings from sensitive and delicate surfaces (138).

Gels with up to 80% of polar, organic solvents can be formed from 40-, 45-, and 75-PVAc with a degree of hydrolysis of 40%, 45% and 75%. However, when the degree of hydrolysis is increased to 80%, the gels are stable only up to a 1:1 mixture of water and organic solvent. The most thermally stable gels are formed when the ratio of the hydroxyl monomer to borate ion is around 15:1.

Fluorescein, cf. Figure 1.24, can be incorporated into the gelated solvent and covalently attached to the polymer chains (139). Fluorescent gels can be used for monitoring the solvent loss from the gel, both peripherally and in the surface layers.

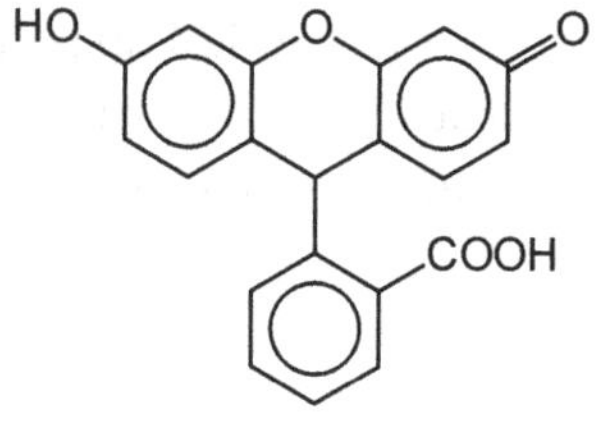

Figure 1.24 Fluorescein.

Gels with 80% ethanol and 20% water have been used to remove the top layer of unbleached shellac from a 19th century Italian wooden frame with a gesso underlayer. The complete removal of the shellac was confirmed by fluorescence analysis. Similarly, the application of these gels to acrylic paint test surfaces was attempted. It seems that gels with large amounts of water may be particularly suitable for this purpose (138).

1.3.13 Restoring Paper Paintings and Calligraphic Works

A method of restoring damaged portions of paper, such as Japanese and machine-made paintings and calligraphic works, has been presented (140). The procedure does not require a high degree of skill, retains the original thickness of the work, improves the durability after restoration and ensures neutralization of the work and simplification of any further restoration.

A raw restoration liquid prepared from paper stock, water and an adhesive, such as a liquid extract of Hibisci radix and optionally a colorant, is placed in a vat. The target paper work to be restored is laid over a cloth or papermaking sheet, which is laid over a filtering sheet-like member that is fitted into a papermaking frame.

The papermaking frame is soaked in the raw restoration liquid so that the fibers contained in the raw restoration liquid are filled into the damaged portions of the paper work. The cloth or papermaking sheet is removed from the papermaking frame while the fibers are dried to the point of dampness. After an additional drying and removal from the papermaking frame the restored paper work is obtained (140).

1.3.13.1 Microblasting for Paper Documents

The use of the microblasting technique with powdered cellulose for dry cleaning documents has been examined (141). Microblasting is a form of abrasive blasting that uses very fine nozzles.

Different cleaning tests were conducted on three documents with different properties. Then, the results were compared to those obtained with traditional dry cleaning with erasers.

In order to assess changes caused to the supports, the treated documents were examined both before and after cleaning with optical and 3D stereomicroscopy and SEM energy dispersive spectrophotometry. The results indicated that microblasting with powdered cellulose is a feasible technique to remove surface dirt or grime on paper documents.

The use of the microblasting technique does not entail changes to the surface properties of the treated supports. Moreover, powdered cellulose is chemically stable and compatible with paper documents, preventing negative long-term effects derived from the presence of rubber residues that may remain in paper fibers when using erasers in dry cleaning treatments (141).

1.3.13.2 *Archaeological Paper Cleaning Using Organic Solvents*

The behavior of organic solvents on the chemical composition of aged paper samples has been evaluated (142). Chemical pulp wood samples have been cleaned with three types of solvents. Then the mechanical properties, color change, and pH changes were checked.

In this way, it could be established if any significant structural or chemical differences between untreated and treated paper is occurring.

Dramatic changes in functional groups on the paper surface, as monitored by FTIR, occurred in the samples before and after solvent cleaning. However, the mechanical properties show that ethyl alcohol, toluene and acetone may give good results in cleaning the paper surface. But these solvents accelerated the oxidation and the hydrolysis of the paper samples under heat aging (142).

1.4 Analysis and Analytical Methods

1.4.1 *Technical Analysis of Paintings*

Technical analysis of paintings is intended to detect retouching, pentimenti, underdrawings and to identify original and added material, e.g., pigments, binding media, coatings, and retouching (143). A pentimento is an alteration of a painting, evidenced by traces of previous work. This shows that the artist has changed his mind in the course of painting (144).

This information is essential for dating and authenticating of the artwork and contributes significantly to our understanding of art objects (143). In addition, it facilitates the evaluation of the physical condition, such as deterioration and intervention, and may be helpful for conservation decisions.

Traditionally, the analysis of paintings has been restricted to invasive investigations and is carried out *ex-situ*. This approach has the drawback of being harmful to the painting. It requires samples to be taken and provides only spot-specific information that is not necessarily representative of the object in its context. Consequently, the development of noninvasive techniques that can be used *in-situ* and provide global information will enjoy a great impact.

Over the last two decades there have been substantial advances made in the application of modern scientific techniques to the chemical and structural analysis of works of art. However, there is still room for improvement as the analysis of artworks is generally a very complex and demanding problem. A very important issue in this respect is the development and use of nondestructive analytical techniques, which can be applied *in-situ*. In response to the demand for nondestructive analytical devices in the art conservation field, a variety of imaging and spectroscopic techniques have been developed and used for the *in-situ* examination of artwork materials.

Film photography has been extensively used to capture artwork images in the visible, near ultraviolet and near infrared parts of the spectrum. A variety of different imaging devices have also been used, ranging from film to analog and digital cameras. Moreover, different cameras with spectral sensitivities restricted either to the visible or to the near infrared part of the spectrum have been used in order to obtain diagnostic information for the artwork under analysis. In particular, imaging in the near infrared enables the visualization of underdrawings, relying on the phenomenon that in general the overlying pigments become transparent in this wavelength range. Broadband fluorescence photography provides information for the condition of coatings, such as varnishes, and enables the localization of previous restoration interventions. A variety of light dispersion spectroscopic techniques have also been used *in-situ* and *ex-situ* for the identification of painting materials. Materials with the same color appearance, determined by the similar diffuse reflectance spectra in the visible, may have different spectral patterns outside the visible part of the spectrum. Compositional alterations associated with the material deterioration can be recorded by measuring absorption, fluorescence, or elastic scattering and nonelastic scattering signals, providing quantitative diagnostic information. Fluorescence and diffuse reflectance spectra are in general broad in the visible and the near infrared part of the spectrum due to the complicated nature of the artwork material and due to light-material interaction mechanisms that are involved in these phenomena. Owing to this fact, these spectroscopic techniques are suitable for detecting *in-situ* chemical and structural alterations and for the differentiation of pigments with the same color appearance but different chemical composition. Raman and FTIR spectroscopy provide improved, molecular-specific diagnostic information, since the acquired spectra contain fingerprint information for a specific point area of the artwork material under analysis. Until now the use of these techniques have been restricted to the experimental-ex situ analysis of material samples, mainly due to the required complicated instrumentation. Laser-induced breakdown spectroscopy is a novel promising technique for in-situ analysis, it requires, however, laser ablation of a spot area and the subsequent spectroscopic analysis of the created plume. For this reason this technique is considered as minimally invasive and in several cases (e.g., fragile and thin material) it is not applicable. Apart from the above mentioned, the common problem that seriously restricts the applicability of conventional spectroscopic methods to the *in-situ* analysis of artwork is that they provide point information, which is inadequate in cases where complicated materials, characterized by a high spatial variability of their contextual features, are examined. Moreover, the point area under analysis has to be determined by the user, which in several cases is not capable of detecting and focusing his attention in artwork areas that are subjected to alterations. This results in a reduction of the accuracy of these methods due to probing errors (143).

In summary, conventional spectrometers provide a large amount of spectroscopic information about one localized site of the object, whereas conventional broadband imaging provides a modest amount of spectral information (resolution), but for a significant area of the object.

In the field of art conservation, there are applications reported where cameras sensitive in the visible and in the near infrared are filtered with optical filters, thus enabling the selection of the imaging center wavelength with the aid of a filter tuning mechanism. In the visible part of the spectrum these cameras are used for accurate color reproduction, while in the near infrared part of the spectrum filter tuning enables the determination of the appropriate imaging band, at which the maximum imaging information for the underlying features is obtained. Based on the above mentioned, it is reasonable to suggest that the combination of the advantages of both imaging and spectroscopy will constitute a significant step forward in nondestructive analysis and documentation of art objects and monuments. Although Raman and FTIR spectroscopy methods provide improved analytical information they are very difficult. In contrast, there is not any fundamental or technological restriction for the development of imaging systems capable of capturing diffuse reflectance and/or fluorescence spectroscopic information for the entire surface under examination. Of course, as mentioned above, these techniques suffer from the main drawback that the captured spectra contain pure information or painting material identification (143).

The comparison of diffuse reflectance and fluorescence spectra of painting material models of known composition has been proposed (143). The intensities emitted and captured at the same wavelengths and for any spatial point of the art object of unknown composition are compared. This comparison, performed for any spatial point of the area of interest, notably improves the diagnostic information and enables the analysis of heterogeneous art objects. This is an imaging method for the nondestructive technical analysis of artistic or historic paintings of unknown structure and composition (143).

1.4.2 *Nondestructive Acoustic Method*

The research related to the detection and determination of the amount of detachment of the separation zones in frescoes is one of the most urgent and difficult problems in the field of conservation and restoration of works of art (145).

Among the nondestructive scientific techniques, the only one which currently is suited to give some information on the presence of separation zones in frescoes, is the so-called thermovision (146).

This technique, however, often gives results which are difficult to interpret. Moreover, its complex instruments, which are not easy to use, and its very high cost, considerably limit the use of this technique (145).

Alternative techniques which could replace thermovision have not reached satisfying results. An example is the ecospectrographic technique. A more recent technique, is based on a vibrometer system. This measures the frequency response function acceleration/force or inertance. Actually, the two above-mentioned acoustic techniques, leaving out of consideration the laboratory results which have been thereby obtained, have the serious limitation that they cannot be classified as nondestructive. In fact, they require that the painting be physically hit by means of a mechanical source in order to excite the structure to be analyzed, and be touched by means of sensors responsible for the detection of the acoustic response.

A different acoustic technique suggested uses the acoustic energy absorption coefficient as the physical indicator of the separation zones. This is a perfectly nondestructive method for the analysis of a painting.

Here, the acoustic source used to excite the surface, and the sensor detecting the acoustic signal reflected by the same, are both positioned at an appropriate distance from the fresco, without needing to touch it. In addition, the proposed method utilizes an appropriate signal processing system allowing the discrimination of feeble signals indicating the detachment zone, from the other noise.

The results were as follows: The specimens simulating the detachments behave like elastic diaphragms or membranes capable of vibrating according to certain vibration modes, and to absorb, for certain frequencies, part of the acoustic energy impinging thereon. The acoustic energy absorption coefficient has revealed itself to be the physical indicator which is most appropriate for the detection of the detachment (145).

1.4.3 Surface Characteristics of Paint

Paint samples were studied to identify the morphological and physical characteristics of the surface before and after cleaning, with and without an intentional abrading. Various analytical methods were used to study the surface morphology, such as (147):

- Digital photography,
- Colorimetry,
- Roughness measurements,
- Stereoscopic microscopy, and
- Scanning electron microscopy.

The results were used to evaluate and categorize the characteristics that contribute to a painting.

It was found that abrasions through the surface of naturally aged discolored paint, such as lead white, lead tin yellow, or copper resinate green, can often be readily observed as points or areas of bright color, usually located on the peaks of impasto. The bright area is then surrounded by an aged, less colorful paint region.

Once familiar with this type of abrasion where contrast is high, a restorer can recognize more subtle shifts in color in broader planes of seemingly intact paint.

The microanalysis confirms that the color shift is actually a result of some differences in the surface structure between an undamaged and a damaged paint (147).

Carefully retouching such an actual damage to the paint surface, with only a slight wisp of retouching paint to eliminate the visual disturbances, can have a considerable effect in unifying the overall appearance of the picture (147).

1.4.3.1 *pH of the Cleaning Solutions*

Poly(acrylic acid) (PAA) thickeners, such as Primal ASE-60, a sodium salt of PAA, are added to acrylics in order to modify the consistency of the paint. Acrylic paints fresh out of the tube are formulated to have a pH of around 9.5, assuring that the PAA acid groups are deprotonated and fully elongated, forming a three-dimensional structure that gives the acrylic paint the creamy texture expected by artists (148). When the paint dries, the PAA hardens but retains the potential to respond to the pH of an aqueous cleaning system. In an alkaline, or even in a neutral cleaning environment, the polymer will hydrate and swell, softening and disrupting the surface of the paint film (52).

In order to minimize the swelling of PAA thickeners, the pH of the cleaning system has to be kept as low as possible.

The choice of buffer is based on the acid dissociation constant, i.e., the pK_a. Actually, the Modular Cleaning Program provides the conservator with information on the effective buffer range of a given weak acid or base.

After the conservator chooses the neutralizing base or acid to adjust the pH of the buffer solution, the Modular Cleaning Program makes a first approximation calculation of the amount of the neutralizing component needed for addition to get the desired pH. Since the calculation is a first approximation, the final pH of all of the stock solutions in the Modular Cleaning Program need to be checked and adjusted with the use of a calibrated pH meter (52).

1.4.4 Binding Media and Protective Coatings

The first reported analytical studies and technical examinations of art and archaeological objects were conducted in the late 18th century. Then, analytical techniques and methods applied to the study of artworks have constantly grown. Among the materials composing the art object, organic compounds used as binding media or protective coatings have attracted attention due to their ability to undergo morphological and chemical changes upon aging. The most recent advances in the identification and determination of organic compounds present in art and art conservation materials have been reviewed (111).

Immunofluorescence techniques have been proposed in recent decades as an alternative to the classical and more simple microchemical tests.

In addition, a variety of instrumental techniques have been improved in an attempt to enhance the sensitivity, repeatability and accuracy of the analytical results.

Spectroscopic techniques, such as UV-vis, FTIR and Raman spectroscopy, have been coupled with light microscopes for these purposes. Synchrotron radiation FTIR microspectroscopy has also been successfully applied to the analysis of artworks. MS has been increasingly used as a detector system coupled with a chromatographic device. The chromatographic methods have also been improved in recent years. Paper and thin layer chromatographic techniques have been progressively replaced by GC, pyrolysis GC, high-performance liquid chromatography and capillary electrophoresis.

More complex proteomics hyphenated techniques, such as nano-liquid chromatography nano-electrospray ionization/collision quadrupole time-of-flight tandem mass spectrometry, have been recently applied to the identification and determination of proteinaceous binders.

Also, microbeam analytical techniques have been used as advanced instrumental techniques for art conservation purposes.

Finally, a number of new instrumental techniques have been proposed as a suitable alternative to the conventional microscopy techniques for morphological studies (111).

1.4.5 *Degradation of Films of Dammar Resin*

Films of dammar resin were aged by their exposure to light and heat (149). The chemical changes in the films were followed by ultraviolet-visible light transmission, infrared and fluorescence spectrophotometry, and gel permeation chromatography. Also, the solubility and the acid number of the films were investigated.

An extensive autoxidation occurred only in the presence of light, although a limited yellowing took place under those conditions.

Autoxidation leads to the formation of products containing polar groups, particularly carboxylic acid groups. Thus, the oxidized resin is soluble only in relatively polar solvents. Crosslinking reactions result in products with a higher molecular weight. However, this does not cause the formation of a significant insoluble fraction.

Yellowing is a secondary non-oxidative thermal process which takes place among autoxidation products. It is particularly strong when heat aging follows aging under light. In the course of heat aging, yellowing is not inhibited when an antioxidant is added or when air is replaced by nitrogen during this step.

Fluorescence under ultraviolet light arises mostly due to thermal non-oxidative reactions. Aldol condensation among the carbonyl species and dehydration reactions are responsible for both yellowing and fluorescence. However, yellow films can be easily bleached with light of wavelengths greater than 400 *nm* (149).

Although films of dammar resin cannot be stabilized against photochemical degradation in the presence of ultraviolet light with currently available additives, a remarkable stabilization is obtained with a HALS if the ultraviolet component of the light source is eliminated (150). Even in the absence of UV light the effect of the HALS can be improved by further addition of a benzotriazole, cf., Figure 1.25, ultraviolet absorber.

It has been proposed that the UV absorber acts as a radical scavenger under those conditions. Films of dammar containing 2% Tinuvin 292 and 3% Tinuvin 328 remain intact when aged in a xenon arc fadeometer under exclusion of UV light for as long as 3,700 *h*, while they normally break down in a few hundred hours. After such a period of accelerated aging the original components of the resin are all still intact and the films remain removable with cyclohexane. No yellowing occurs during heat aging after a preceding period of prolonged fadeometer aging (150).

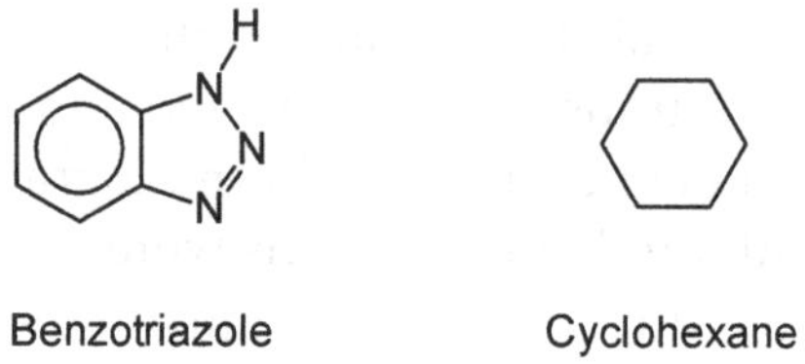

Figure 1.25 Benzotriazole and cyclohexane.

1.4.6 *Spectroscopic Techniques*

The application of selected spectroscopic techniques to the field of cultural heritage conservation has been reviewed (151).

In recent times, science and technology have been applied to the world of cultural heritage, conservation and preservation. In this field, spectroscopic techniques represent one of the most powerful tools to investigate the structure of all the materials constituting a cultural object.

In particular, some spectroscopic techniques are important for the characterization of the following works of art:

- Easel paintings,
- Paper and ink artifacts,
- Stone monuments,
- Wall paintings, and
- Archaeological ceramics.

The spectroscopic techniques that have been used are Fourier transform infrared spectroscopy which is also coupled with a microscope, X-ray fluorescence spectroscopy, proton-induced X-ray emission spectroscopy, ultraviolet, visible and fluorescence spectroscopy, Raman spectroscopy, and Mössbauer spectroscopy.

1.4.6.1 *Dye Characterization*

Dyes are among the most significant components in works of art and archaeological findings. In the scientific examination of historical artefacts, the identification of natural dyestuffs is a challenging task, due to the complexity of their chemical composition and the possible presence of mixtures of chromophores and degradation products (152). For this reason, new analytical procedures and techniques have been developed and improved for the characterization of organic dyes and their identification in microsamples.

Several analytical methods based on spectrometric and chromatographic techniques that have been reviewed are important for the study of organic dyes in works of art and archaeological findings (152). These are, among others, molecular spectroscopy, high-performance liquid chromatography, and mass spectrometry.

1.4.6.2 *Spectral Imaging Technology*

Spectral imaging technology, which simultaneously records spectral and spatial information about an object, was initially developed for remote sensing. Since then it has been successfully applied to other areas of research. Actually, this noninvasive method of characterization has provided promising results in the analysis of paintings and written documents, the characterization of materials and digital documentation (153).

The literature dealing with the application of multispectral and hyperspectral imaging for the study and conservation of works of art have been reviewed (153).

Illuminated manuscripts are complex multilayered and multimaterial objects which include a support, preparatory layer, pigment layers, and often highlights.

For example, two Italian illuminations from the treasury of the Saint-André Cathedral in Bordeaux were studied with hyperspectral imaging (154). The characteristics of the spectral signal in the visible range have been used to map pigments as the reflectance spectra were compared with a reference library of medieval pigments. In the two Italian miniatures, the palette could be identified and was found to correspond to typical medieval pigments, such as lapis lazuli, red lead, lead white, green copper-based pigment, and probably anthraquinone-based pigments, such as brazilwood and kermes (154).

1.4.7 *Organic Mass Spectroscopy*

The issues of organic mass spectroscopy in art and archaeology have been collected in a monograph (155). An overview of the analysis of art and archaeological materials using techniques based on mass spectrometry is presented there. Also, the basic principles, procedures and applications of mass spectrometric techniques are illustrated. In this way, this monograph fills a gap in the field of application on destructive methods in the analysis of museum objects.

Pyrolysis GC has become a powerful technique for the characterization of nonvolatile materials (156, 157). It has been widely used in forensic science and industrial laboratories, but its application to conservation has been limited, due to the compositional complexity and small sample size of the materials available for examination.

Furthermore, the absence of a reliable compound reference library for identification of major materials in artworks has restricted comparative studies on actual artifacts. The use of Pyrolysis GC for the general study of materials, e.g., drying oils, amber, and natural resins, has been reviewed and specific examples have been given to provide an introduction to applications of this technique in art and archaeology (156).

Rosin, together with polysaccharide gums, have been identified by pyrolysis GC/MS in Egyptian cartonnages that date back to 2000 BC (158).

1.4.8 *Portable NMR*

The use of a mobile device based on a nuclear magnetic resonance spectroscopy (NMR) single-sided sensor for *in-situ* noninvasive determination of the moisture content of wood items has been proposed, especially for items of cultural heritage interest (159).

The moisture content can be obtained through the moisture volume fraction, which is an appropriate quantity for the sensor and corresponds to the fraction of its measurements sensitive volume occupied by water. The device has been used to assess changes in the moisture content of wood specimens caused by changes over time of the environmental relative humidity. The kinetics of water adsorption has been related to results obtained with the gravimetric method (159).

Measurements on an old painting, the Pietà (1516–1517), an oil on poplar wood panel by Sebastiano del Piombo (1485 Venice, 1547 Rome), Civic Museum, Viterbo, Italy, have conclusively shown the good sensitivity of the sensor and its capability to behave as a non-invasive and *in-situ* utilizable device.

The results of *in-situ* painting measurements showed that the NMR sensor can track moisture fluctuations that are outside the sensitivity range and precision of electro-hygrometric approach (159).

1.4.8.1 NMR Mouse

The durability of historic wall paintings is highly dependent on environmental influences such as moisture ingress, salt crystallization and temperature changes (160).

A fundamental understanding of dynamic transport processes in wall paintings is necessary to apply suitable conservation and restoration methods to preserve such objects with high cultural value. Noninvasive, mobile NMR techniques with single-sided sensors enable the monitoring of the moisture content, transport and apparent diffusion constants in wall paintings. This technique was investigated by experiment and also by modeling to correlate salt crystallization, moisture transport and local diffusion in wall painting samples (160).

Moreover, the influence of different painting techniques, both fresco and secco, and conservation and consolidation methods on moisture transport and diffusion have been discussed. The results were also compared with the results from field measurements on real fresco paintings in Casa del Salone Nero and the Villa of the Papyri, Herculaneum, Italy (160).

1.5 Forgeries

There are monographs dealing with forgeries of paintings (161–166).

Also, chemical methods that can be used for the detection of forgeries have been reviewed (167).

1.5.1 *Image Analysis Tools*

Whether machine learning and image analysis tools can be used to assist art experts in the authentication of unknown or disputed paintings has been examined (168). The reexamination of some of recently successful experiments showed that variations in image clarity in the experimental data sets could be correlated with authenticity, and may have acted as a confounding factor, artificially improving the results.

To determine the extent of this factor's influence on previous results, a new ground truth data set has been provided, in which originals and copies are known and the image acquisition conditions are uniform. Multiple previously successful methods are found to be ineffective on this new confounding factor-free data set, but it could be demonstrated that supervised machine learning on features derived from hidden Markov tree modeling of the paintings' wavelet coefficients has the potential to distinguish copies from originals in the new data set (168).

1.5.2 *Correlation Filters*

Statistical and image processing techniques have been described that can enhance the existing capabilities of analyzing works of art for authenticity (169). The applicability of advanced correlation filters for identifying forgeries among disputed paintings has been discussed.

By training the optimal trade-off synthetic discriminant function filter on each section of a coarsely parceled image of an original painting, it is possible to distinguish between a low-quality digitized representation of a painting and its forgery. The methods also specifically indicate where the differences occur and where the replica is particularly faithful to the original. This method is also valuable in determining whether an original painting has undergone any modifications, given that a representation of the initial version is available (169).

1.5.3 *X-Ray Analysis*

Electromagnetic radiation is commonly used for medical diagnosis, and also sometimes to treat diseases. Medicine, however, is not the only industry which uses electromagnetic radiation methods. Electromagnetic radiation methods have a long-standing role in the forensic examination of cultural properties (170).

The roles of X-ray methods, γ-ray methods, and β particles for the detection of forgery and guiding the restoration of paintings have been discussed. X-ray techniques play the largest role in painting authentication and restoration. The information that can be obtained from the different technologies in the forensic examination of paintings has been discussed.

Currently, the best method which uses electromagnetic radiation to examine paintings in 2- or 3D, is optical coherence tomography. Optical coherence tomography is not a widely known modality to radiologists, it is a known technique in ophthalmology. The physics of optical coherence tomography and the information yielded during painting examination have been discussed (170).

1.5.4 Contourlet Transform

A method based on the contourlet transform (171) for the purpose of art authentication has been presented (172). Here sample digital images from different artists are converted into grayscale images. Then, the background patches from the grayscale images are extracted.

The extracted patches are transformed using the method of contourlet transform and modeled using hidden Markov tree. The Fisher distance information is calculated to measure the stylistic similarity between the brushwork of the artist. The Fisher distance is a measure of the dissimilarity between two probability distribution functions (173).

The result obtained in this way was analyzed to determine the authenticity of the paintings. Using this method, different current existing solutions have been analyzed to determine their limitations.

A new hybrid method has been proposed that can overcome some limitations. The method has been tested on samples of twenty images of the paintings, consisting of the paintings from Van Gogh and other artists separated as Van Gogh and non Van Gogh paintings (172).

References

1. S. Hackney, The art and science of cleaning paintings, in M.F. Mecklenburg, A.E. Charola, and R.J. Koestler, eds., *New Insights into the Cleaning of Paintings*, Vol. 3 of *Smithsonian Contributions to Museum Conservation*, p. 11, Washington, D.C., 2013. Universidad Politécnica de Valencia and Museum Conservation Institute, Smithsonian Institution Scholarly Press.
2. M. Doerner, *Malmaterial und seine Verwendung im Bilde*, Verlag für Praktische Kunstwissenschaft, München, 6th edition, 1921, Reprint 1938.
3. C. Sitwell, *Studies in the History of Painting Restoration*, Archetype Publications in association with the National Trust, London, 1998.
4. R. Wolbers, *Cleaning Painted Surfaces: Aqueous Methods*, Archetype Publications, London, 2000.
5. W. Percival-Prescott, *Studies in Conservation*, Vol. 35, p. 73, September 1990.
6. T. Caley, *Studies in Conservation*, Vol. 35, p. 70, September 1990.
7. R. Wolbers, A. Norbutus, and A. Lagalante, Cleaning of acrylic emulsion paints: Preliminary extractive studies with two commercial paint systems, in M.F. Mecklenburg, A.E. Charola, and R.J. Koestler, eds., *New Insights into the Cleaning of Paintings*, Vol. 3 of *Smithsonian Contributions to Museum Conservation*, pp. 147–158, Washington, D.C., 2013. Universidad Politécnica de Valencia and Museum Conservation Institute, Smithsonian Institution Scholarly Press.
8. K. Laudenbacher, Considerations of the cleaning of paintings, in M.F. Mecklenburg, A.E. Charola, and R.J. Koestler, eds., *New Insights into the Cleaning of Paintings*, Vol. 3 of *Smithsonian Contributions to Museum Conservation*, p. 7, Washington, D.C., 2013. Universidad Politécnica de Valencia and Museum Conservation Institute, Smithsonian Institution Scholarly Press.
9. J. Clifton, *The Conservation and Restoration of Paintings: An Introduction*, McFarland & Co., 1988.
10. C.M. Groen, *Paintings in the Laboratory: Scientific Examination for Art History and Conservation.* Ph.D thesis, Amsterdam Institute for Humanities Research, Amsterdam, 2011.
11. S.G. Fernández-Villa, *Ge-conservación/conservação*, Vol. 8, p. 58, 2015.
12. A. Massing, *Painting Restoration Before La Restauration: The Origins of the Profession in France*, Harvey Miller Hamilton Kerr Institute, University of Cambridge, London, Cambridge, 2012.
13. S. Schmitt, *Zeitschrift für Kunsttechnologie und Konservierung*, Vol. 4, p. 30, 1990.

14. K. Sutherland, B. Price, A. Lins, I. Passeri, M. Mecklenburg, A. Charola, and R. Koestler, Oxalate-rich surface layers on paintings: Implications for interpretation and cleaning, in M.F. Mecklenburg, A.E. Charola, and R.J. Koestler, eds., *New Insights into the Cleaning of Paintings*, Vol. 3 of *Smithsonian Contributions to Museum Conservation*, pp. 85–88, Washington, D.C., 2013. Universidad Politécnica de Valencia and Museum Conservation Institute, Smithsonian Institution Scholarly Press.
15. J. Stoner, *The Conservation of Easel Paintings*, Routledge, London, 2012.
16. R. Drewello and R. Weissmann, *Applied Microbiology and Biotechnology*, Vol. 47, p. 337, April 1997.
17. C. Messal, T. Gerber, and G. Ballin, *Materials and Corrosion*, Vol. 50, p. 166, 1999.
18. V. Alunno-Rossetti and M. Marabelli, *Studies in Conservation*, Vol. 21, p. 161, November 1976.
19. J. Scully, K. Nassau, P.K. Gallagher, A.E. Miller, and T.E. Graedel, *Corrosion Science*, Vol. 27, p. 669, 1987.
20. L.S. Selwyn, N.E. Binnie, J. Poitras, M.E. Laver, and D.A. Downham, *Studies in Conservation*, Vol. 41, p. 205, January 1996.
21. M. Cotte, P. Walter, G. Tsoucaris, and P. Dumas, *Vibrational Spectroscopy*, Vol. 38, p. 159, 2005.
22. R. Arbizzani and U. Casellato, The Scrovegni Chapel at Padova: Chemical investigations on the polychromy of wooden furnishings, in *The Fifth Infrared and Raman Users Group Meeting*, pp. 117–119, Los Angeles, 2002. Getty Conservation Institute.
23. K. Sutherland, B.A. Price, I. Passeri, and M. Tucker, A study of the materials of Pontormo's *Portrait of Alessandro de Medici*, in *Symposium OO – Materials Issues in Art and Archaeology VII*, Vol. 852 of *MRS Proceedings*, 2004.
24. C. Higgitt and R. White, *National Gallery Technical Bulletin*, Vol. 26, p. 89, 2005.
25. A. van Loon, *Color Changes and Chemical Reactivity in Seventeenth-century Oil Paintings*. Ph.D thesis, Swammerdam Institute for Life Sciences (SILS), Amsterdam, 2008.
26. M. Matteini, A. Moles, G. Lanterna, C. Lalli, M.R. Nepoti, M. Rizzi, and I. Tosini, *Characteristics of the materials and techniques* in M. Ciatti and M. Seidel, eds., *Giotto: The Santa Maria Novella Crucifix*, pp. 387–403. EDIFIR-Edizioni Firenze, Florence, 2002.
27. A. Casoli, Z. Di Diego, and C. Isca, *Environ. Sci. Pollut. Res.*, Vol. 21, p. 13252, March 2014.
28. M. Daudin-Schotte, M. Bisschoff, I. Joosten, H. van Keulen, and K.J. van den Berg, Dry cleaning approaches for unvarnished paint surfaces, in M.F. Mecklenburg, A.E. Charola, and R.J. Koestler, eds.,

New Insights into the Cleaning of Paintings, Vol. 3 of *Smithsonian Contributions to Museum Conservation*, pp. 209–219, Washington, D.C., 2013. Universidad Politécnica de Valencia and Museum Conservation Institute, Smithsonian Institution Scholarly Press.

29. E. Estabrook, *Journal of the American Institute for Conservation*, Vol. 28, p. 79, 1989.
30. A.W. Brokerhof, S. de Groot, L.J. Pedersoli, Jr., H. van Keulen, B. Reissland, and F. Ligterink, *Papierrestaurierung*, Vol. 3, p. 13, 2002.
31. M. Noehles, *Papierrestaurierung*, Vol. 3, p. 22, 2002.
32. S. Schorbach, *Zeitschrift für Kunsttechnologie und Konservierung*, Vol. 23, p. 41, 2009.
33. A. Phenix and K. Sutherland, *Reviews in Conservation*, Vol. 2, p. 47, 2001.
34. A. Phenix, *Zeitschrift für Kunsttechnologie und Konservierung*, Vol. 12, p. 387, 1998.
35. A.F.M. Barton, *Chem. Rev.*, Vol. 75, p. 731, December 1975.
36. A. Phenix, *WAAC Newsletter*, Vol. 20, September 1998.
37. J.L. Gardon and J.P. Teas, *Solubility parameters* in R.R. Myers and J.S. Long, eds., *Treatise on Coatings Part II*, Vol. 2, pp. 413–471. Marcel Dekker, New York, 1976.
38. G.D. Smith and R. Johnson, *WAAC Newletter*, Vol. 30, p. 11, January 2008.
39. C. Stavroudis and S. Blank, *WAAC Newsletter*, Vol. 11, p. 2, May 1989.
40. C. McGlinchey, *WAAC Newsletter*, Vol. 24, p. 17, May 2002.
41. A. Phenix, *Journal of the American Institute for Conservation*, Vol. 41, January 2002.
42. T. Fardi, E. Stefanis, C. Panayiotou, S. Abbott, and S. van Loon, *Journal of Cultural Heritage*, Vol. 15, p. 583, November 2014.
43. S. Zumbühl, E.S.B. Ferreira, N.C. Scherrer, and V. Schaible, The non-ideal action of binary solvent mixtures on oil and alkyd paint: Influence of selective solvation and cavitation energy, in M.F. Mecklenburg, A.E. Charola, and R.J. Koestler, eds., *New Insights into the Cleaning of Paintings*, Vol. 3 of *Smithsonian Contributions to Museum Conservation*, pp. 97–105, Washington, D.C., 2013. Universidad Politécnica de Valencia and Museum Conservation Institute, Smithsonian Institution Scholarly Press.
44. Y. Marcus, *Solvent Mixtures: Properties and Selective Solvation*, Marcel Dekker, Inc., New York, 2002.
45. C. Reichardt, *Solvents and Solvent Effects in Organic Chemistry*, VCH, Weinheim, Germany, New York, NY, 1988.
46. J. Chipperfield, *Non-Aqueous Solvents*, Oxford University Press, Oxford, New York, 1999.

47. K.-H. Hellwege, *Thermodynamic equilibria of boiling mixtures* in J. Weishaupt, ed., *Macroscopic and Technical Properties of Matter*, Vol. 3 of *Landolt Börnstein – Numerical Data and Functional Relationships in Science and Technology*, pp. 1–276. Springer, Berlin, 1975.
48. J.A.L. Domingues, N. Bonelli, R. Giorgi, E. Fratini, F. Gorel, and P. Baglioni, *Langmuir*, Vol. 29, p. 2746, 2013.
49. E. Carretti, L. Dei, R.G. Weiss, and P. Baglioni, *Journal of Cultural Heritage*, Vol. 9, p. 386, September–December 2008.
50. R. Giorgi, M. Baglioni, D. Berti, and P. Baglioni, *Acc. Chem. Res.*, Vol. 43, p. 695, June 2010.
51. D. Chelazzi, G. Poggi, Y. Jaidar, N. Toccafondi, R. Giorgi, and P. Baglioni, *Journal of Colloid and Interface Science*, Vol. 392, p. 42, February 2013.
52. C. Stavroudis and T. Doherty, The modular cleaning program in practice: Application to acrylic paintings, in M.F. Mecklenburg, A.E. Charola, and R.J. Koestler, eds., *New Insights into the Cleaning of Paintings*, Vol. 3 of *Smithsonian Contributions to Museum Conservation*, p. 139, Washington, D.C., 2013. Universidad Politécnica de Valencia and Museum Conservation Institute, Smithsonian Institution Scholarly Press.
53. C. Stavroudis and T. Doherty, *WAAC Newsletter*, Vol. 29, p. 9, 2007.
54. A. Murray, C.C.d. Berenfeld, S.S. Chang, E. Jablonski, T. Klein, M.C. Riggs, E.C. Robertson, and W.A. Tse, The condition and cleaning of acrylic emulsion paintings, in *Symposium II – Materials Issues in Art and Archaeology VI*, Vol. 712 of *MRS Proceedings*, p. II1.4 (8 pages), 2002.
55. J.R. Druzik and G.R. Cass, A new look at soiling of contemporary paintings by soot in art museums, in *The Indoor Air Quality Meeting for Museums Conference Report*, pp. 22–27, 2000.
56. B. Ormsby, P. Smithen, F. Hoogland, C. Miliani, and T. Learner, A scientific evaluation of surface cleaning acrylic emulsion paintings, in J. Bridgeland, ed., *Preprints ICOM-CC 15th Trienn Conf New Delhi*, pp. 865–873, 2008.
57. B. Ormsby and T. Learner, *Studies in Conservation*, Vol. 54, p. 29, 2009.
58. S. Croll, T.J.S. Learner, and P. Smithen, Overview of developments in the paint industry since 1930, in *Modern Paints Uncovered: Proceedings from the Modern Paints Uncovered Symposium*, pp. 17–29. Getty Publications, 2007.
59. S.F. Zumbühl, N. Attanasio, W. Scherrer, N. Müller, N. Fenners, and W. Caseri, Solvent action on dispersion paint systems and the influence on the morphology – changes and destruction of the latex microstructure, in T. Learner, P. Smithen, J. Krüger, and M. Schilling, eds., *Modern Paints Uncovered*, pp. 257–268, Los Angeles, 2007. The Getty Conservation Institute.

60. A. Pereira, M.J. Melo, P. Eaton, S. Schäfer, and T. Learner, A preliminary study into the effects of cleaning polyvinyl acetate paints, in M.F. Mecklenburg, A.E. Charola, and R.J. Koestler, eds., *New Insights into the Cleaning of Paintings*, Vol. 3 of *Smithsonian Contributions to Museum Conservation*, p. 135, Washington, D.C., 2013. Universidad Politécnica de Valencia and Museum Conservation Institute, Smithsonian Institution Scholarly Press.
61. E. Signorini, Surface cleaning of paintings and polychrome objects in Italy: The last 15 years, in M.F. Mecklenburg, A.E. Charola, and R.J. Koestler, eds., *New Insights into the Cleaning of Paintings*, Vol. 3 of *Smithsonian Contributions to Museum Conservation*, p. 17, Washington, D.C., 2013. Universidad Politécnica de Valencia and Museum Conservation Institute, Smithsonian Institution Scholarly Press.
62. D.L. Weaire, *The Physics of Foams*, Clarendon Press, Oxford, New York, 1999.
63. A. Heckenbücker and P. Demuth, Surface cleaning with aqueous foams, in M.F. Mecklenburg, A.E. Charola, and R.J. Koestler, eds., *New Insights into the Cleaning of Paintings*, Vol. 3 of *Smithsonian Contributions to Museum Conservation*, pp. 225–227, Washington, D.C., 2013. Universidad Politécnica de Valencia and Museum Conservation Institute, Smithsonian Institution Scholarly Press.
64. J.L. Kaplan and R. Namiat, Method to remove foxing stains from paper & celluloid items, US Patent 7 462 202, December 9, 2008.
65. I. Hausmann and P. Demuth, The Schlürfer: A vacuum technique for the cleaning of paintings, in M.F. Mecklenburg, A.E. Charola, and R.J. Koestler, eds., *New Insights into the Cleaning of Paintings*, Vol. 3 of *Smithsonian Contributions to Museum Conservation*, pp. 221–224, Washington, D.C., 2013. Universidad Politécnica de Valencia and Museum Conservation Institute, Smithsonian Institution Scholarly Press.
66. J. Brus and P. Kotlík, *Studies in Conservation*, Vol. 41, p. 109, 1996.
67. J. Podany, K.M. Garland, W.R. Freeman, and J. Rogers, *Journal of the American Institute for Conservation*, Vol. 40, p. 15, Spring 2001.
68. S. Chapman and D. Mason, *Journal of The American Institute for Conservation*, Vol. 42, p. 381, 2003.
69. S.Q. Lomax and S.L. Fisher, *Journal of the American Institute for Conservation*, Vol. 29, p. 181, Autumn 1990.
70. E.R. de la Rie and A.M. Shedrinsky, *Studies in Conservation*, Vol. 34, p. 9, 1989.
71. J. Arslanoglu and T. Learner, *The Conservator*, Vol. 25, p. 62, 2001.
72. P. Spathis, E. Karagiannidou, and A.-E. Magoula, *Studies in Conservation*, Vol. 48, p. 57, 2003.
73. E. Carretti, L. Dei, and P. Baglioni, *Langmuir*, Vol. 19, p. 7867, 2003.

74. E. Carretti, B. Salvadori, P. Baglioni, and L. Dei, *Studies in Conservation*, Vol. 50, p. 128, 2005.
75. R. Bellucci, P. Cremonesi, and G. Pignagnoli, *Studies in Conservation*, Vol. 44, p. 278, 1999.
76. P. Pouli, A. Nevin, A. Andreotti, P. Colombini, S. Georgiou, and C. Fotakis, *Applied Surface Science*, Vol. 255, p. 4955, February 2009.
77. M. Cooper, *Laser Cleaning in Conservation: An Introduction*, Butterworth-Heinemann, Woburn, MA, 1998.
78. C. Fotakis, D. Anglos, S. Georgiou, V. Tornari, and V. Zafiropulos, *Lasers in the Preservation of Cultural Heritage: Principles and Applications*, Vol. 2 of *Series in Optics and Optoelectronics*, Taylor and Francis, 2006.
79. S. Georgiou, V. Zafiropulos, D. Anglos, C. Balas, V. Tornari, and C. Fotakis, *Applied Surface Science*, Vol. 127-129, p. 738, 1998.
80. M. Lassithiotaki, A. Athanassiou, D. Anglos, S. Georgiou, and C. Fotakis, *Applied Physics A: Materials Science and Processing*, Vol. 69, p. 363, 1999.
81. M. Gómez-Heras, M. Alvarez de Buergo, E. Rebollar, M. Oujja, M. Castillejo, and R. Fort, *Applied Surface Science*, Vol. 219, p. 290, 2003.
82. M.L. Wolbarsht and A. deCruz, Method for cleaning artwork, US Patent 5 951 778, assigned to deCruz and Wolbarsht, September 14, 1999.
83. Wikipedia, Er:yag laser — wikipedia, the free encyclopedia, 2016. [Online; accessed 20-April-2016].
84. D. Thompson, *The Materials and Techniques of Medieval Painting*. Dover Publications, New York, 1956.
85. Wikipedia, Gesso — wikipedia, the free encyclopedia, 2016. [Online; accessed 1-February-2017].
86. B.A. Banks and S.K. Rutledge, Process for non-contact removal of organic coatings from the surface of paintings, US Patent 5 560 781, assigned to The United States of America as represented by the Administrator of the National Aeronautics and Space Administration, October 1, 1996.
87. P. Cremonesi, Rigid gels and enzyme cleaning, in M.F. Mecklenburg, A.E. Charola, and R.J. Koestler, eds., *New Insights into the Cleaning of Paintings*, Vol. 3 of *Smithsonian Contributions to Museum Conservation*, pp. 179–183, Washington, D.C., 2013. Universidad Politécnica de Valencia and Museum Conservation Institute, Smithsonian Institution Scholarly Press.
88. B. Veríssimo Mendes, K.J. Berg, L. Megens, I. Joosten, and M. Daudin, *New approaches to surface cleaning of unvarnished contemporary oil paintings – moist sponges and cloths* in J.K. van den Berg, A. Burnstock,

M. de Keijzer, J. Krueger, T. Learner, A. Tagle, de, and G. Heydenreich, eds., *Issues in Contemporary Oil Paint*, pp. 373–388. Springer International Publishing, Cham, 2014.

89. M. Baglioni, R. Giorgi, D. Berti, and P. Baglioni, *Nanoscale*, Vol. 4, p. 42, 2012.
90. A. Nualart-Torroja, M. Oriola-Folch, and M. Mascarella-Vilageliu, Cleaning issues for nine Francesc Artigau pop art paintings, in M.F. Mecklenburg, A.E. Charola, and R.J. Koestler, eds., *New Insights into the Cleaning of Paintings*, Vol. 3 of *Smithsonian Contributions to Museum Conservation*, pp. 175–178, Washington, D.C., 2013. Universidad Politécnica de Valencia and Museum Conservation Institute, Smithsonian Institution Scholarly Press.
91. M. Pons, *Conservation Issues in Modern and Contemporary Murals*, Cambridge Scholars Publishing, Newcastle upon Tyne, UK, 2015.
92. E.R. de la Rie, *Analytical Chemistry*, Vol. 61, p. 1228A, 1989.
93. C.A. Maines and E.R. de la Rie, *Progress in Organic Coatings*, Vol. 52, p. 39, January 2005.
94. E.R. de la Rie and C.W. McGlinchey, *Studies in Conservation*, Vol. 35, p. 168, 1990.
95. I. Bonaduce, M.P. Colombini, I. Degano, F. Di Girolamo, J. La Nasa, F. Modugno, and S. Orsini, *Analytical and Bioanalytical Chemistry*, Vol. 405, p. 1047, 2013.
96. H. Piena, *Journal of the American Institute for Conservation*, Vol. 40, p. 59, 2001.
97. V. Routledge, *WAAC Newsletters*, Vol. 22, p. N.A., May 2000. electronic: http://cool.conservation-us.org/waac/wn/wn22/wn22-2/wn22-206.html.
98. M.F. Pacheco, A.I. Pereira, A.J. Parola, and L.C. Branco, The use of ionic liquids for varnish removal: Effectiveness and risk evaluation, in M.F. Mecklenburg, A.E. Charola, and R.J. Koestler, eds., *New Insights into the Cleaning of Paintings*, Vol. 3 of *Smithsonian Contributions to Museum Conservation*, pp. 233–236, Washington, D.C., 2013. Universidad Politécnica de Valencia and Museum Conservation Institute, Smithsonian Institution Scholarly Press.
99. L.C. Branco, J.N. Rosa, J.J. Moura Ramos, and C.A.M. Afonso, *Chemistry - A European Journal*, Vol. 8, p. 3671, 2002.
100. C.F. Poole, *Journal of Chromatography A*, Vol. 1037, p. 49, May 2004.
101. J. Dupont, *Ionic liquids: Structure, properties and major applications in extraction/reaction techonology* in C.A.M. Afonso and J.P.G. Crespo, eds., *Green Separation Processes: Fundamentals and Applications*, pp. 229–245. John Wiley, New York, 2005.
102. K. Sutherland, *Studies in Conservation*, Vol. 45, p. 54, 2000.

103. J. Romero-Noguera, M.d.M. López-Miras, I. Martín-Sánchez, J.M. Ramos-López, and F. Bolívar-Galiano, *Electronic Journal of Biotechnology*, Vol. 13, p. 0, November 2010.
104. R. Morrison, *National Gallery Technical Bulletin*, Vol. 31, p. 112, 2010.
105. G.A. Berger, *Formulating adhesives for the conservation of paintings* in N.S. Brommelle and P. Smith, eds., *Conservation and Restoration of Pictorial Art*. Butterworths-Heinemann, London, 1976.
106. C.V. Horie, *Materials for Conservation: Organic Consolidants, Adhesives and Coatings*, Conservation and Museology Series, Elsevier, Butterworth-Heinemann, Oxford, reprint edition, 2005.
107. B.A. Ormsby, A. Soldano, M.H. Keefe, A. Phenix, and T. Learner, *AIC Paint Spec. Gr. Postprints*, Vol. 23, p. 77, 2010.
108. H.I. Zeliger and F.J. Nussbaum, Method of restoring and conserving oil paintings and resulting integrated permanent structure, US Patent 5 019 441, assigned to Zeliger and Nussbaum, May 28, 1991.
109. A. Barros, *Adhesives and Consolidants in Painting Conservation*, Archetype Publications, in association with the ICON Paintings Group, London, 2012.
110. R.E. Fieux, Means and method of restoring documents, paintings and the like, US Patent 4 330 586, May 18, 1982.
111. M.T. Doménech-Carbó, *Analytica Chimica Acta*, Vol. 621, p. 109, July 2008.
112. J. Mills and R. White, *The Organic Chemistry of Museum Objects*, Butterworth-Heinemann, Oxford England Boston, 1994.
113. C.V. Horie, *Materials for Conservation: Organic Consolidants, Adhesives and Coatings*, Routledge, New York, 2010.
114. J. Nadolny, *Studies in Conservation*, Vol. 48, p. 39, 2003.
115. A. Eibner, *Mouseion*, Vol. 13/14, p. 70, 1931.
116. W. Ostwald, *Sitzungsberichte der Königlich Preußischen Akademie der Wissenschaften*, Vol. 5, 1905.
117. J.L. Ferreira, M.J. Melo, and A.M. Ramos, *Polymer Degradation and Stability*, Vol. 95, p. 453, April 2010.
118. A. Roche, *Studies in Conservation*, Vol. 41, p. 45, 1996.
119. A. Cataldi, A. Dorigato, F. Deflorian, and A. Pegoretti, *Polymer Engineering & Science*, Vol. 55, p. 1349, March 2015.
120. P. Baglioni, *Nanoscience for the Conservation of Works of Art*, Royal Society of Chemistry, Cambridge, 2013.
121. E. Carretti, E. Fratini, D. Berti, L. Dei, and P. Baglioni, *Angew. Chem. Int. Ed.*, Vol. 48, p. 8966, November 2009.
122. I. Natali, M.L. Saladino, F. Andriulo, D. Chillura Martino, E. Caponetti, E. Carretti, and L. Dei, *Journal of Cultural Heritage*, Vol. 15, p. 151, March 2014.

123. P. Baglioni, D. Berti, M. Bonini, E. Carretti, L. Dei, E. Fratini, and R. Giorgi, *Advances in Colloid and Interface Science*, Vol. 205, p. 361, 2014.
124. P.G. De Gennes and C. Taupin, *J. Phys. Chem.*, Vol. 86, p. 2294, June 1982.
125. P. Baglioni, D. Chelazzi, R. Giorgi, and G. Poggi, *Langmuir*, Vol. 29, p. 5110, 2013.
126. E. Carretti, R. Giorgi, D. Berti, and P. Baglioni, *Langmuir*, Vol. 23, p. 6396, May 2007.
127. R. Giorgi, L. Dei, and P. Baglioni, *Studies in Conservation*, Vol. 45, p. 154, September 2000.
128. M. Ambrosi, L. Dei, R. Giorgi, C. Neto, and P. Baglioni, *Langmuir*, Vol. 17, p. 4251, July 2001.
129. V. Daniele, G. Taglieri, and R. Quaresima, *Journal of Cultural Heritage*, Vol. 9, p. 294 , 2008.
130. B. Salvadori and L. Dei, *Langmuir*, Vol. 17, p. 2371, 2001.
131. C. Rodriguez-Navarro, I. Vettori, and E. Ruiz-Agudo, *Langmuir*, Vol. 32, p. 5183, 2016.
132. ASTM International, Test methods for measuring adhesion by tape test, ASTM Standard ASTM D3359-97, ASTM International, West Conshohocken, PA, 2009.
133. C.A. Bishop, *Adhesion and adhesion tests* in C.A. Bishop, ed., *Vacuum Deposition onto Webs, Films and Foils*, chapter 9, pp. 197–208. William Andrew Publishing, Boston, 3rd edition, 2015.
134. P. Koneczny, Paint conservation-restoration composition and method, US Patent Application 20 080 262 165, October 23, 2008.
135. F. Valentini, A. Diamanti, M. Carbone, E.M. Bauer, and G. Palleschi, *Applied Surface Science*, Vol. 258, p. 5965, 2012.
136. M. Rizzo, L. Machado, S. Borrely, M. Sampa, P. Rela, J. Farah, and R. Schumacher, *Radiation Physics and Chemistry*, Vol. 63, p. 259, March 2002.
137. S.W. Sinton, *Macromolecules*, Vol. 20, p. 2430, October 1987.
138. L.V. Angelova, B.H. Berrie, and R.G. Weiss, Partially hydrolyzed polyvinyl acetate and borax gels for cleaning painted surfaces, in M.F. Mecklenburg, A.E. Charola, and R.J. Koestler, eds., *New Insights into the Cleaning of Paintings*, Vol. 3 of *Smithsonian Contributions to Museum Conservation*, pp. 201–203, Washington, D.C., 2013. Universidad Politécnica de Valencia and Museum Conservation Institute, Smithsonian Institution Scholarly Press.
139. X. Guan, X. Liu, Z. Su, and P. Liu, *Reactive and Functional Polymers*, Vol. 66, p. 1227, 2006.
140. M. Kihara, Method of restoring paper paintings and calligraphic works, US Patent 5 958 506, September 28, 1999.

141. M. Ángel Iglesias-Campos, C. Ruiz-Recasens, and E. Diaz-Gonzalez, *Journal of Cultural Heritage*, Vol. 15, p. 365, 2014.
142. R.R.A. Hassan, *Mediterranean Archaeology and Archaeometry*, Vol. 15, p. 141, 2015.
143. K. Balas and D. Pelecoudas, Imaging method and apparatus for the non-destructive analysis of paintings and monuments, US Patent 7 042 567, assigned to Foundation of Research and Technology (Crete, GR), May 9, 2006.
144. Wikipedia, Pentimento — wikipedia, the free encyclopedia, 2017. [Online; accessed 1-February-2017].
145. G.B. Cannelli and P. Calicchia, Nondestructive acoustic method and device, for the determination of detachments of mural paintings, US Patent 6 728 661, assigned to Consiglio Nazionale Delle Ricerche (Rome, IT), April 27, 2004.
146. R.S. Sharpe, *Research Techniques in Nondestructive Testing*, Academic Press, London, New York, 1970.
147. L. Sozzani, A preliminary investigation into the surface characteristics of paint and the implications for restoration, in M.F. Mecklenburg, A.E. Charola, and R.J. Koestler, eds., *New Insights into the Cleaning of Paintings*, Vol. 3 of *Smithsonian Contributions to Museum Conservation*, p. 23, Washington, D.C., 2013. Universidad Politécnica de Valencia and Museum Conservation Institute, Smithsonian Institution Scholarly Press.
148. T. Learner, *Analysis of Modern Paints*, Getty Conservation Institute, Los Angeles, 2004.
149. E.R. de la Rie, *Studies in Conservation*, Vol. 33, p. 53, 1988.
150. E.R. de la Rie and C.W. McGlinchey, *Studies in Conservation*, Vol. 34, p. 137, 1989.
151. G. Bitossi, R. Giorgi, M. Mauro, B. Salvadori, and L. Dei, *Applied Spectroscopy Reviews*, Vol. 40, p. 187, 2005.
152. I. Degano, E. Ribechini, F. Modugno, and M.P. Colombini, *Applied Spectroscopy Reviews*, Vol. 44, p. 363, 2009.
153. C. Fischer and I. Kakoulli, *Studies in Conservation*, Vol. 51, p. 3, 2006.
154. A. Mounier and F. Daniel, *Studies in Conservation*, Vol. 60, p. S200, 2015.
155. M.P. Colombini and F. Modugno, eds., *Organic Mass Spectrometry in Art and Archaeology*, Wiley, Chichester, West Sussex, 2009.
156. A. Shedrinsky, T.P. Wampler, N. Indictor, and N.S. Baer, *Conservation of Cultural Property in India*, Vol. 21, p. 35, 1988.
157. A. Shedrinsky and N.S. Baer, *The application of analytical pyrolysis to the study of cultural materials* in T. Wampler, ed., *Applied Pyrolysis Handbook*, chapter 6, pp. 105–132. CRC Press/Taylor & Francis, Boca Raton, 2nd edition, 2007.

158. D. Scalarone and O. Chiantore, *Py-GC/MS of natural and synthetic resins* in M.P. Colombini and F. Modugno, eds., *Organic Mass Spectrometry in Art and Archaeology*, chapter 12, pp. 327–362. Wiley, Chichester, West Sussex, 2009.
159. L. Senni, C. Casieri, A. Bovino, M.C. Gaetani, and F. De Luca, *Wood Science and Technology*, Vol. 43, p. 167, 2009.
160. D. Oligschläger, S. Waldow, A. Haber, W. Zia, and B. Blümich, *Magnetic Resonance in Chemistry*, Vol. 53, p. 48, 2015.
161. R.G. Reisner, *Fakes and Forgeries in the Fine Arts: A Bibliography*, Special Libraries Association, New York, 1950.
162. S. Schüller, *Forgers, Dealers, Experts: Strange Chapters in the History of Art*, Putnam, 1960.
163. W. Fong, *Artibus Asiae*, Vol. 25, p. 95, 1962.
164. L.G.G. Savage, *Forgeries, Fakes and Reproductions. A Handbook for the Collector*, London, 1963.
165. R.H. Marijnissen, *Paintings: Genuine-Fraud-Fake*, Elsevier, 1985.
166. S.A. Cole, *Connoisseurship all the way down: Art authentication, forgery, fingerprint identification, expert knowledge* in N. Charney, ed., *Art Crime: Terrorists, Tomb Raiders, Forgers and Thieves*, chapter 3, pp. 27–32. Palgrave Macmillan UK, London, 2016.
167. F.E. Rogers, *Journal of Chemical Education*, Vol. 49, p. 418, 1972.
168. G. Polatkan, S. Jafarpour, A. Brasoveanu, S. Hughes, and I. Daubechies, Detection of forgery in paintings using supervised learning, in *2009 16th IEEE International Conference on Image Processing (ICIP)*, pp. 2921–2924, Nov 2009.
169. P. Buchana, I. Cazan, M. Diaz-Granados, F. Juefei-Xu, and M. Savvides, Simultaneous forgery identification and localization in paintings using advanced correlation filters, in *2016 IEEE International Conference on Image Processing (ICIP)*, pp. 146–150, Sept 2016.
170. T.C. Moran, A.D. Kaye, A. Rao, and F.R. Bueno, *Journal of Forensic Radiology and Imaging*, Vol. 5, p. 38, 2016.
171. Wikipedia, Contourlet — wikipedia, the free encyclopedia, 2016. [Online; accessed 30-January-2017].
172. R. Karki, A. Alsadoon, P.W.C. Prasad, A.M.S. Rahma, and A. Elchouemi, A proposed method on painting authentication using contourelet transform (pauct), in *2016 7th International Conference on Information and Communication Systems (ICICS)*, pp. 213–216, April 2016.
173. S.I.R. Costa, S.A. Santos, and J.E. Strapasson, *Discrete Applied Mathematics*, Vol. 197, p. 59, 2015. Distance Geometry and Applications.

2

Textiles

The chemical principles of textile conservation have been described in a monograph (1).

2.1 Textile Colors

2.1.1 Historical Development of Colorants

Color belongs to the elements of nature that make human life more aesthetic and fascinating in the world (2). Plants, animals, and minerals have been used as primary sources for colorants, dyes or pigments since ancient times.

The first fiber dyes that have been known in prehistoric times consisted of fugitive stains from berries, blossoms, barks, and roots. A historical introduction to colorants has been presented. The chronological developments in dyeing processes have been detailed and the origins of dyes used (2).

Scale insects are small insects of the order Hemiptera. They comprise the superfamily Coccoidea. Most scale insects are parasites of plants (3).

Scale insects have long been used to produce crimson-colored dyes. Many plant pigments are used as dyes. Madder was utilized long ago as a red vegetable dye for leather, wool, cotton and silk. Indigo is a natural dye, which is structurally related to betalains. The ancient dye Tyrian purple, has been extracted from shellfish of the *Murex* genus (2). New discoveries about the science of color have resulted in many industrial innovations and sharp changes in fashion.

2.1.2 *Classification of the Used Colorants*

Due to degradation, the few textiles that have been recovered in eastern North America are discolored brown or black. Also, some textiles are green due to mineralization (4). The original coloration remained unrecognized.

A protocol has been developed to classify the colorants used in these materials that is minimally destructive (4). The method includes forensic photography, optical microscopy of strategically sampled fibers, scanning electron microscopy and energy dispersive spectrometry.

Because of ultraviolet reflection, fluorescence and infrared photography capture differences due to the chemical composition, which cannot necessarily be seen by the unaided eye and the purposive sampling is facilitated by the protocol.

The observation with an optical microscope may offer a thoroughly saturated color in the fibers that is indicative of dyeing in comparison to a deposited pigment on the surface. The elemental composition of the fibers and the surface particulate can be correlated with potential mineral or organic colorant sources, as well as reflecting a fiber mineralization (4).

2.1.3 *Microanalysis of Organic Pigments in Ancient Textiles*

New methods based on surface-enhanced Raman scattering (SERS) have been reviewed (5). This method is used for the nondestructive and minimally invasive identification of organic colorants in certain objects such as historic and archaeological textiles, paintings, and drawings.

An ecocompatible homogeneous nanostructured agar matrix was used for extraction of the components. The extraction system was adapted according to the chemical properties of the target analyte by choosing appropriate reagents for the extraction and optimizing the extraction time.

The developed methodology for the selective extraction and identification of anthraquinones and indigoids in the typical concentration used in textiles has been described. Anthraquinonic dyes, like alizarin and carminic acid, are usually bound to textile fibers by a mordant like alum that introduces Al^{3+} ions. These ions act as an ionic bridge between the fiber and the dye molecule. The bonding mechanism of alizarin on textiles is shown in Figure 2.1.

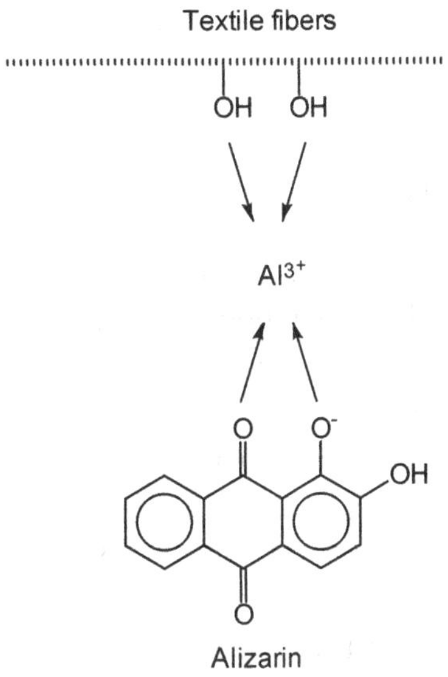

Figure 2.1 Bonding mechanism of alizarin on textiles (5).

The molecules can be tentatively transferred from the textile fiber to an agar gel either by using a solvent in which they are highly soluble, e.g., an alcohol, or by detaching them from the fiber by specific chemical processes.

Initially, a few ethyl alcohol droplets were added into the agar gel and then contacted with the fiber for some 30 *min*. Another method was used with ethylenediamine tetraacetic acid (EDTA). This chelating agent is able to detach the dye by a selective removal of the Al^{3+} ions from the system. The use of EDTA acid results in an acidic system with a pH of around 3. The use of its disodium salt results in solutions at pH values almost neutral.

It was found that the system was extremely stable, easy to use and produce, and easy to store. Also it was found that the samples could be analyzed even after long time intervals, maintaining its enhancement properties unaltered, without a detriment to the extracted compounds (5).

Highly structured SERS band intensities have been obtained from the extracted dyes adopting laser light excitations at 514.5 *nm* and 785 *nm* of a micro-Raman system.

This analytical method has been found to be extremely safe for the analyzed substrates, thus being a promising procedure for the selective analysis and detection of molecules at low concentrations in the field of art conservation (5).

2.1.4 *Analysis of Dyes*

The analysis of dyes from three textile specimens associated with human remains found in the Chehrabad salt mine in northwestern Iran dating to 2000–400 years BP has been described (6). All of the specimens analyzed were made from tabby woven wool.

These dyes are unique in this part of the world, not only because of their age, but they also represent textiles used by common people, i.e., salt miners, as opposed to funerary garments of the wealthy people at that time.

Samples of yarns from these specimens were extracted and analyzed using high-performance liquid chromatography (HPLC) with a diode array detector and mass spectroscopy (MS). It was found that the red dye was obtained from madder (*Rubia tinctorum* L.). The blue dye was obtained from an indigo plant, probably woad (*Isatis tinctoria* L.). Both of these plants are known in Iran.

Also, two yellow, plant-derived, flavonol dyes were found. The first flavonol dye seems to come from a species of tamarisk (*Tamarix* sp.). The second flavonol dye, was found in both yellow and green yarns. It is until now an unidentified plant (6).

2.1.4.1 *Chemical Analysis of Murex-Dyed Textiles*

The chemical and chromatographic analysis techniques have been reported that were used in the study of three textiles that were found in a cave in Wadi Murabba'at, in the Judean desert (7).

These textiles date to the Roman period and were dyed with a prestigious purple dye. HPLC identified the *Murex* sea snail *Hexaplex trunculus* as a source of the dye for the three textiles. However, the process of dyeing was not uniform. The color of one textile was modified to greenish-blue, probably by the exposure to the sun during the dyeing process. The two other textiles were subject to a double-dyeing process, and underwent a second dipping in a dye solution extracted from the Armenian cochineal insect. This process resulted in a reddish-purple color, which was indicative of high status (7).

2.1.4.2 *Extraction and Analysis of Madder Dyes*

A new alkaline mild extraction technique of madder dyes in lake and textile has been presented (8). This technique is based on the use of ammonia with disodium EDTA. This method was developed to preserve the glycoside compounds fixed on the fiber.

For testing, samples of dyed textile and lake were used that were prepared according to ancient recipes. The extracts from these samples were analyzed and characterized by 1D and 2D nuclear magnetic resonance spectroscopy diffusion-ordered spectroscopy and electrospray ionization MS.

Afterwards the method was evaluated using real samples that were analyzed by HPLC diode array detection and HPLC MS. It was confirmed that the ammonia extracts of madder dyed yarn and lake are mainly composed of glycosylated anthraquinones, as described in the literature (8).

2.1.5 *Organic Residue Analysis*

An analysis was conducted on pottery from a Middle Bronze Age workshop from the Minoan culture of Crete, confirming it as a facility for manufacturing organic dyes (9). As analytical technique, gas chromatography (GC) was used.

The archaeological site under investigation is located at Alatzomouri-Pefka in the northeast of Crete. The site has a series of nine rock-cut basins associated with channels in the bedrock, remains of stone walls, and many artifacts. The pottery is dated in Middle Minoan IIB (ca. 1800/1700 BC).

The GC studies could identify three dyes (*Murex* purple, yellow from weld, and red from madder) as well as lanolin, the oil associated with wool from sheep (9).

2.1.6 *Infrared Analysis*

Infrared spectra of the particulate material shed from fragile archaeological textiles have been compared with the spectra of known materials in order to classify them by their composition (10,11).

The spectra were collected of comparative plant and animal fibers, including charred, mineralized, pigmented and dyed examples. Further, spectra of particulates from fabrics from two archaeological sites in eastern North America were collected.

A separation of the comparative materials into groups based on their composition could be achieved, as well as some distinctions being made in dyed, pigmented, and mineralized samples. Some compositional features of the archaeological materials could be discerned that can be useful for their identification (11).

2.1.6.1 *Reflection Fourier Transform Mid-Infrared Spectroscopy*

Silk embroideries and cotton grounds of ancient Caucasian (Kaitag) textiles were analyzed *in-situ* by a portable Fourier transform infrared spectrometer equipped with a reflection module (12). Differently colored areas were analyzed in order to identify the dyes fixed on the fibers.

The so-obtained spectra were elaborated by calculating the corresponding second derivative, and a library search was then performed using a database including the second derivative spectra of a large range of historical dyes and the corresponding undyed fibers. Some of the detected colors are summarized in Table 2.1.

These results suggest that this technique, combined with the library search method, has a good capability of recognizing natural dyes on both types of ancient textile fibers, and this takes place in an entirely nondestructive way (12).

Table 2.1 Detected colors (12).

Century	Color	Ingredients
17th	Red Cotton	Madder Purpurin Alizarin Carmine Ellagic acid
17th	Dark green Silk	Indigo Gamboge Tannic acid Logwood Kamala
17th	Dark brown Silk	Tannic acid Saffron Gallic acid Sumac, aqueous extract Sumac, methanol extract
17th	Blue Cotton	Indigo Kamala Purpurin Buckthorn lake Madder
18th	Red Silk	Alizarin Madder Walnut Tannic acid Kamala
18th	Dark blue Silk	Indigo Saffron Henna Safflower Alkanet
18th	Black Cotton	Ellagic acid Brazilwood Logwood Saffron Indigo

2.2 Textiles from Various Locations

2.2.1 *Early Textiles and Textile Production in Europe*

The occurrence of textiles and the textile production in Europe from prehistory to AD 400 have been reviewed (13).

Also, the power of textiles in late antiquity have been reviewed in a monograph (14). A lot of examples of textiles have been depicted there. Ideals of character and beauty, and conceptions of self and society, were in flux during Late Antiquity, a period of extensive dramatic cultural upheaval for the Roman world, as the extraordinary growth of Christianity eclipsed paganism. Textiles from Late Antiquity document transformations of the cultural traditions and the societal value of the contemporary population.

2.2.1.1 *Greek and Roman Textiles and Dress*

The current research concerning the study of textiles and dress in classical antiquity has been reviewed in a monograph. The special issues that have been addressed are (15):

1. The importance of studying textiles to understand the economy and landscape of the past,
2. Different types of embellishments of dress from weaving techniques of embroidery,
3. The close links between the language of ancient mathematics and weaving,
4. The relationships of iconography to the realities of clothed bodies, including a paper on the groundbreaking research on the polychromy of ancient statuary; dye recipes and methods of analysis,
5. Case studies of garments in Spanish, Viennese and Greek collections which discuss the methods of analysis and conservation,
6. Analysis of textile tools from across the Mediterranean, and
7. Discussions of trade and ethnicity regarding the workshop relations in Roman fulleries.

2.2.1.2 *Textiles of Northern Mediterranean*

The role of textiles in the northern Mediterranean region has been reviewed (16). In the past, textile manufacturing has been practiced at all levels of society. This was one of the most labor-intensive industries. As such, it was an industry of great cultural and social importance, playing an integral part in local and regional economies as well as local, regional, and long-distance exchange.

The social significance of textile production in the northern Mediterranean was also expressed in funerary rituals by the inclusion of textile implements together with burial goods; as well as in religious activities, through the deposition of textile tools in votive deposits and special ritual cloth production in sanctuaries (16).

2.2.2 *Natural Organic Dyes from Ancient Europe*

Textile research has become an important field of archaeology (17). Many dyeing techniques which were used during prehistoric times are still in use today. An experimental study was done to reproduce some natural organic dye processes from ancient times, in order to better interpret the archaeological textile material.

Natural organic dyes can be classified in various ways. However, there are three main categories (17):

1. Direct dyes,
2. Mordant dyes, and
3. vat dyes.

Direct dyes are obtained in a one-bath process. These dyes are usually water-soluble and bind to the material themselves. Such dyes deteriorate with time because the binding forces are quite weak. Alkaloid dyes and carotenoid dyes, such as orchid and saffron, belong to this group (17).

On the other hand, vat dyes are water insoluble and have to be chemically altered to a soluble and often colorless intermediate. The dyes yield a very stable color when the soluble form is bound to the fabric. These dyes need certain additional ingredients to make them colorfast, i.e., a base and a reducing agent. The use of vat dyes can be traced to ancient times with some of the earliest references going back thousands of years to when Tyrian purple was used. This was of animal origin, namely an extract from shellfish. Indigo was the most important of the natural coloring materials in this group, as well as woad and Tyrian purple (17).

Most natural organic dyes require the addition of a mordant instead of a leveling agent, and these are referred to as mordant dyes. Basic mordants are made from metal salts, such as aluminium, copper, iron, or tin. Dyes that belong to different color categories may need different types of a mordant, such as aluminium for madder and iron, or tin for a purple color (17).

2.2.3 *Ancient Liturgical Vestment*

Yarns of an ancient liturgical vestment (chasuble) have been analyzed using inductively coupled plasma optical emission spectrometry (18). A chasuble is the outermost liturgical vestment worn by priests for the celebration of the holy mass.

The samples were analyzed with respect to Al, Ag, Au, Cd, Co, Cr, Cu, Ni, Pb, and Zn using amounts of 0.1–1 *mg*.

With the exception of a single sample, Ag was found to be the most abundant element in all the yarns.

In the samples that were containing silver, it ranged from 68 to 97%. Only two samples contained aluminum.

The chasubles have been reported to be an artifact of the 15th century; however, a part of the yarns is composed of materials that have been attributed to a later period. Actually, the use of aluminum in metal alloys arrived only in later periods (18).

2.2.4 *Textiles and Dyes in Pre-Columbian Northern Chile*

As in most ancient and modern cultures, clothing and textiles carried great psychosocial meaning to the people in pre-Columbian South America (19). In order to understand the importance of colors and dyestuffs in these ancient societies 765 samples from 256 textiles have been analyzed. The samples originate from the Museo Arqueológico San Miguel de Azapa of the Universidad de Tarapacá in northern Chile.

The samples ranged from the Chinchorro to the Inca cultures (ca. 8000 BCE to 1500 CE). A wider range of dyes was used in the region than expected. Annatto, cochineal and relbunium, red dyes well known from textiles elsewhere in the Andes, were used next to three unidentified red dyes.

Although outside influence appeared to have had some impact on dye choice, local residents continued to use traditional dyes throughout the prehistory of the region, even in times of increased regional interaction. A significant difference in the use of various dyes between coastal and more inland sites was observed, especially during the Late Intermediate and Late Horizon periods (ca. 1000 to 1532 CE), thus indicating increasing differences in social status and the access to dyestuffs (19).

2.2.5 *Painted Andean Textiles*

Many Andean textiles are painted, a tradition practiced across the Andes as early as the Lithic period (ca. 10000 to 4000 BC) (20). Samples from nine painted Andean textiles were collected from different museums and analyzed at the Getty Conservation Institute in Los Angeles. The samples represent a range of cultures, locations, and times. As analytical techniques polarized light microscopy and X-ray fluorescence spectroscopy were used.

Because of the paucity of published data on painted textiles, a comparison with mural paintings has been also made. The results revealed that Andean painters used a range of pigments, including ocher and cinnabar. The pigments were chosen based on their chemical and physical characteristics from at least the Early Horizon period (ca. 950 BC to 200 CE) onward, although there were also cultural preferences for specific pigments.

The quality of the cinnabar pigment especially shows that Andean painters had great technical knowledge and skills, superior to what was previously suspected. Patterns of pigment choice and application indicate that the painted textiles were primarily produced for funerary purposes (20).

2.2.6 *Textiles from the Silk Road*

The collection of textiles from Xiongnu burial sites was obtained in recent years as a result of the research of the Russian-Mongolian expedition led by N. Polosmak (21).

Throughout their long history, Xiongnu controlled the Central Asia regions of the Silk Road, by which many products, including textiles and wool, were brought from the West to China. Woolen fabrics and textiles of high quality were found in the Xiongnu noble burial sites located in the mountains of Mongolia.

The analysis of the composition of these dyes by HPLC showed that the woolen fabrics were dyed with plant dyestuffs and also with insect dyestuffs. Each analyzed sample was dyed with a set of dyestuffs, indicating that dyers not only had the necessary and different dyes, but possessed highly developed craftsmanship of dyeing (21).

The results indicate that the dyeing of woolen textiles found in the graves of the Xiongnu nobility was carried out in the manufactories of the Mediterranean region, known for their fabric dyeing culture. Numerous Chinese-made silk fabrics were dyed with traditional Han epoch plant dyes, i.e., indigo and Indian madder.

The composition of the dyes of silk textiles fundamentally differs from dyes of the woolen fabrics by the absence of dyestuffs of insect origin (21).

2.2.6.1 *Dyes in Ancient Textiles from Yingpan, Xinjiang*

Natural dyes have been analyzed in well-preserved, ancient textiles from Yingpan, an important archaeological site on the Silk Road (22). HPLC with a diode array detector and MS was used for analysis. The dyes were removed from the textile fibers using a soft extraction method. Here, pyridine, water and oxalic acid were used for extraction.

Two species of madder-type plants, *Rubia tinctorum* and *Rubia cordifolia* were identified as the dye sources in red and brown samples.

Also, unusual flavonoid glycosides in yellow and green woolen yarns were possibly found. However, these compounds could not be accurately associated with a specific plant that may have been used to produce these dyes (22).

However, it has been found that yellow and green silk threads have been colored with protoberberine-type dyes, probably obtained from species of cork tree, *Phellodendron* spp.

It has been proposed that the textile dyes found at Yingpan reflect a merging of Eastern and Western practices and technologies during the 800-year period (206 BCE to 618 CE), thus encompassing the early years of the Silk Road (22).

2.2.7 *Historical Chinese Dyestuffs*

Ultra-high performance liquid chromatography (UHPLC) coupled with both photodiode array detection and electrospray ionization MS was used for the chemical characterization of common textile dyes in ancient China.

Three different extraction methods were tested: Extraction with a mixture of dimethyl sulfoxide and oxalic acid, extraction with dimethyl sulfoxide, and extraction with hydrochloric acid.

Dimethyl sulfoxide is especially suitable for dyes that bind to the textile fiber via hydrophobic interactions, i.e., vat dyes and direct dyes. On the other hand, oxalic acid and hydrochloric acid can be used for breaking the metal-dye bonds of mordant dyes.

The usage of hydrochloric acid is considered to be a harsh method. The acid breaks down the fabric-dye and the metal-dye bonds very effectively. But in the course of application the sugar bonds are also destroyed. In addition, some components are chemically changed by hydrolysis, decarboxylation, and esterification reactions.

Several common dyes used in ancient China were investigated. Most of the dyes are of reliable botanical provenance (23). These materials were used to prepare reference dyed samples for the use of a database for the analysis method used here. The materials used are collected in Table 2.2.

Table 2.2 Dyes used for a data base (24).

Material	Latin name
Safflower	*Carthamus tinctorius* L., root
Sappanwood	*Caesalpinia sappan* L., heartwood
Gromwell	*Lithospermum* sp., root
Chinese cork tree	*Phellodendron chinense* Schneid., bark
Turmeric	*Curcuma longa* L., rhizome
Pagoda bud	*Styphnolobium japonicum* L.
Gardenia	*Gardenia jasminoindes* f. *longicarpa*
Indigo	*Strobilanthes cusia* (Nees)
Indigofera	*Indigofera tinctoria* L.
Dyer's woad	*Isatis tinctoria* L., leaf
Gallnut	Produced by insect *Melaphis chinensis*
Acorn cup	*Quercus acutissima*
Munjeet	*Rubia cordifolia* L.
Smoketree	*Cotinus coggygria* var. chinerea

The phenomena of esterification and isomerization of the dye constituents of gallnut, gardenia and saffron, and the dye composition of acorn cup dyed silk, could be clarified.

Using UHPLC coupled with both photodiode array detection and electrospray ionization MS as method of analysis, some characteristic components that have not been reported earlier on dyed textiles could be clarified, especially for the dyes gallnut, acorn cup, gardenia and munjeet. Furthermpre, components of similar chemical structures showed improved separation and could be clearly identified. The chromatographic, UV-vis spectral and mass spectral data of the characteristic components of the dyes have been presented in a very detailed way in the original paper (24).

Also, 6-hydroxyrubiadin and its glycosides were first reported on a dyed sample with *Rubia cordifolia* from China. 6-Hydroxyrubiadin is shown in Figure 2.2.

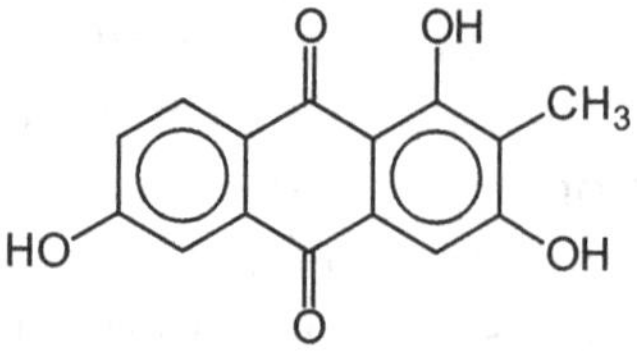

Figure 2.2 6-Hydroxyrubiadin.

In summary, the research results dealt with above form an important foundation for the identification and interpretation of dyes on historical and archaeological Chinese textiles and textiles from other geographical areas where similar dyes were used (24).

2.2.8 *Ancient Indonesian Textiles*

The characteristics of ancient Indonesian textiles, in particular sacred cloths, have been investigated (25). The analysis revealed that animism developed in various ways in Indonesia from 500 BC to AD 1800. In this region, Buddhism, Hinduism, and Islam were active. The Indonesian regions can be divided into six big regions that produce textiles. These are the islands of Sumatra, Borneo, Java, Celebes, Nusa Tenggara, and Bali.

By space and time, the colors of Indonesian textiles represent the ideology of one religion. Indonesia produces primary colors of red, yellow, and blue. The colors have been produced by extracting leaves of Indigo, Indian Mulberry root shell, Sappanwood branches, Candlenut fruit, Turmeric root, and Mangosteen rind (25).

2.3 Processing Methods

2.3.1 *Ancient Chemical Processing of Organic Dyes and Pigments*

The dyer in ancient periods was an advanced empirical chemist (26). While inorganic pigments produced magnificent colors, the most elaborate chemical processing of colorants in antiquity, from the source to the final product, involved organic dyes from flora and fauna sources. There, the dyers applied their practical chemical knowledge to botany, entomology, and malacology.

By controlling temperature and alkaline or acidic pH of the dye bath, the dyers were able to create a colorful textile by dyeing. These dyes could survive even some six millennia. The various chemical principles that were applied by an ancient master of colorful chemistry have been discussed (26).

2.3.2 *Color Preservation of Ancient Natural Dyes*

The reasons why ancient natural dyes, such as indigo, preserve their colors and why other natural dyes, such as brazilein, seem to degrade much faster, have been elucidated (27).

Studies in photochemistry can answer many important questions about the stability of materials. Photochemistry explains what happens when light meets molecules. Thus, photochemistry is providing new tools for the conservation of artworks (27).

2.3.3 *Flavonols for Textile Dyeing*

Yellow, flavonol-based dyes have been widely used in the past (28).

This class of natural dyes presents special problems because flavonols (3-hydroxyflavones) can readily undergo photo-oxidation, resulting in fading or discoloration. However, if the 3-hydroxy group is substituted, e.g., with a glycosyl group, the flavonol derivative is relatively stable. 3-Hydroxyflavone is shown in Figure 2.3.

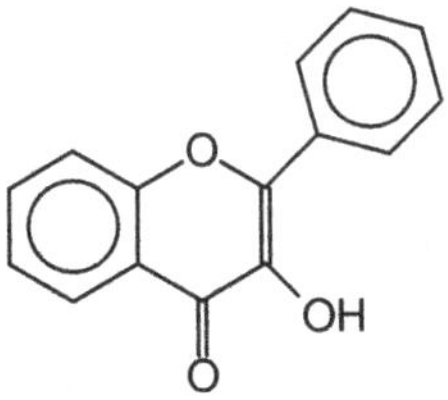

Figure 2.3 3-Hydroxyflavone.

HPLC has been suggested to be useful in identifying the plants used for dyeing ancient textiles. Representatives of eight plant genera have been analyzed: *Delphinium, Flaveria, Pistachia, Prangos,*

Rhamnus, *Sophora*, *Tamarix*, and *Vitis*. The goal was to identify the flavonol components present in these plant genera.

It was found that in all the genera, the principal colorants were flavonol 3–O-glycosides, besides, in one case, 3–O-sulfates. Some plants also contained the light-labile flavonol aglycones.

Also, some strategies are discussed that may have been used in the past to reduce the concentration of flavonol aglycones (28).

References

1. A. Balazsy, *Chemical Principles of Textile Conservation*, Butterworth-Heinemann, Oxford, Boston, 1998.
2. A. Gürses, M. Açıkyıldız, K. Güneş, and M.S. Gürses, *Historical development of colorants* in *Dyes and Pigments*, pp. 1–12. Springer International Publishing, Cham, 2016.
3. Wikipedia, Scale insect — wikipedia, the free encyclopedia, 2017. [Online; accessed 24-January-2017].
4. C.M. Baldia and K.A. Jakes, *Toward the classification of colorants in archaeological textiles of Eastern North America* in M. Glascock, R.J. Speakman, and R.S. Popelka-Filcoff, eds., *Archaeological Chemistry: Analytical Techniques and Archaeological Interpretation*, Vol. 968 of *ACS Symposium Series*, chapter 2, pp. 15–43. American Chemical Society, Distributed by Oxford University Press, Oxford, New York, 2007.
5. M. Ricci, C. Lofrumento, E. Castellucci, and M. Becucci, *Journal of Spectroscopy*, Vol. 2016, p. 1, 2016.
6. C. Mouri, A. Aali, X. Zhang, and R. Laursen, *Heritage Science*, Vol. 2, p. 20, 2014.
7. N. Sukenik, A. Varvak, Z. Amar, and D. Iluz, *Journal of Archaeological Science: Reports*, Vol. 3, p. 565, 2015.
8. L. Lombardi, I. Serafini, M. Guiso, F. Sciubba, and A. Bianco, *Microchemical Journal*, Vol. 126, p. 373, 2016.
9. A.J. Koh, P.P. Betancourt, M.N. Pareja, T.M. Brogan, and V. Apostolakou, *Journal of Archaeological Science: Reports*, Vol. 7, p. 536, 2016.
10. K. Jakes, *Archaeological Chemistry: Materials, Methods, and Meaning*, American Chemical Society, Distributed by Oxford University Press, Washington, DC, New York, 2002.
11. K.A. Jakes, C.M. Baldia, and A.J. Thompson, *Infrared examination of fiber and particulate residues from archaeological textiles* in M. Glascock, R.J. Speakman, and R.S. Popelka-Filcoff, eds., *Archaeological Chemistry: Analytical Techniques and Archaeological Interpretation*, Vol. 968 of *ACS Symposium Series*, chapter 3, pp. 44–77. American Chemical Society, Distributed by Oxford University Press, Oxford, New York, 2007.

12. E. De Luca, S. Bruni, D. Sali, V. Guglielmi, and P. Belloni, *Applied Spectroscopy*, Vol. 69, p. 222, 2015.
13. M. Harlow, *Britannia*, Vol. 45, p. 483, 11 2014.
14. T.K.T. und Jennifer L. Ball, ed., *Designing Identity: The power of Textiles in Late Antiquity*, Princeton University Press, New York, NY Princeton, NJ, 2016.
15. M. Harlow and M.-L. Nosch, eds., *Greek and Roman textiles and Dress: An Interdisciplinary Anthology*, Vol. 19 of *Ancient Textiles Series*, Oxbow Books, Oxford & Philadelphia, 2014.
16. M. Gleba, *Domestic archaeology: Textiles of northern mediterranean* in C. Smith, ed., *Encyclopedia of Global Archaeology*, pp. 2165–2172. Springer, New York, NY, 2014.
17. X.M. Yang and M.L. Wu, Experimental study of natural organic dyes from ancient europe, in *Advances in Textile Engineering and Materials III*, Vol. 821 of *Advanced Materials Research*, pp. 593–596. Trans. Tech. Publications, 12 2013.
18. D. Amorello, S. Barreca, M. Bruno, A. Milia, S. Orecchio, and A. Pettignano, *Microchemical Journal*, Vol. 129, p. 305, 2016.
19. H. Barnard and R. Boytner, Textile dyes in pre-columbian northern chile, in I. Karapanagiotis, G.-T. Stamkopoulos, and N. Loudovikos, eds., *Dyes in History & Archaeology*, DHA 34, p. 19, Thessaloniki, 2015. University Ecclesiastical Academy of Thessaloniki, Department of Management and Conservation of Ecclesiastical Cultural Heritage Objects.
20. H. Barnard, R. Boytner, N. Khandekar, and M. Schleicher, *Ñawpa Pacha*, Vol. 36, p. 209, 2016.
21. E. Karpova, V. Vasiliev, V. Mamatyuk, N. Polosmak, and L. Kundo, *Journal of Archaeological Science*, Vol. 70, p. 15, 2016.
22. J. Liu, C. Mouri, R. Laursen, F. Zhao, Y. Zhou, and W. Li, *Journal of Archaeological Science*, Vol. 40, p. 4444, December 2013.
23. J. Han, *Economic Botany*, Vol. 69, p. 230, September 2015.
24. J. Han, J. Wanrooij, M. van Bommel, and A. Quye, *Journal of Chromatography A*, Vol. 1479, p. 87, 2017.
25. K.-C. Langi and S. Park, *Journal of the Korean Society of Costume*, Vol. 66, p. 67, 2016.
26. Z.C. Koren, *Modern chemistry of the ancient chemical processing of organic dyes and pigments* in S.C. Rasmussen, ed., *Chemical Technology in Antiquity*, chapter 7, pp. 197–217. American Chemical Society, 2015.
27. M.J. Melo, J.L. Ferreira, A.J. Parola, and J.S.S. de Melo, *Photochemistry for cultural heritage* in G. Bergamini and S. Silvi, eds., *Applied Photochemistry: When Light Meets Molecules*, Vol. 92 of *Lecture Notes in Chemistry*, pp. 499–530. Springer International Publishing, Cham, 2016.
28. C. Mouri, V. Mozaffarian, X. Zhang, and R. Laursen, *Dyes and Pigments*, Vol. 100, p. 135, 2014.

3

Archaeological Wood

There are monographs concerning archaeological wood, its properties, chemistry, and preservation (1–4).

The first step in studying archaeological wood is to define how it is preserved, in order to use the most suitable preparation procedure (5). Archaeological wood may be waterlogged, dried, charred, or mineralized. Two commonly used procedures are sectioning and fracturing.

Sectioning is applied to waterlogged or desiccated wood and Fracturing is used with charcoal and mineralized wood. Both preparation techniques for wood and charcoal are fast and effective and allow the analysis of large numbers of wood fragments in a relatively short time.

However, they have the main disadvantage of being destructive techniques, producing further fragmentation and loss of study material. For this reason, it is recommended that the preparation of archaeological woods should be always carried out with caution not to produce irreparable damage to materials that in some cases may be precious and unique (5).

3.1 Analysis Methods

3.1.1 Assessment of Commonly Used Cleaning Methods

The technology, materials and conservation: scanning electron microscope (SEM) and microanalysis has been reviewed (6). Further, a study was conducted to diagnose and evaluate the effect of commonly used cleaning methods in Egypt on the anatomical structure of archaeological wood samples (7). Beech wood samples, which were taken from anonymous mashrabia, have been cleaned mechanically and chemically, then a SEM study was undertaken, to monitor any significant structural changes in wood samples due to cleaning processes.

The SEM data, however, showed that cleaning procedures, both mechanical and chemical, affect the anatomical structure of wood, and do not achieve the best result. The main problem is that the effect of reagents cannot be easily removed from the wood structure. Ethyl alcohol proved to have minimal effect on the wood structure in this study (7).

3.1.2 *Predicting the Wood Preservation Status*

Different types of waterlogged environments involve a variety of natural processes, either in favor of preservation or acting to degrade the organic material (8).

Also, the environmental parameters of most significance for wood preservation may differ even at the same geographical location. These parameters are interrelated, but also act upon the wood and wood-degrading organisms independently.

The conditions and the mechanisms at the interface, such as the sediment-water interface, are of specific interest because of their extreme variability of activities (9).

For an improved understanding of the different biogeochemical mechanisms of most significance for wood preservation, and also to be able to connect them with some key parameters in the wreck site environment, it is essential to have further insight into the seabed sediment interactions at different wreck sites (8). An important approach to investigate the role of the sediments involves the documentation of the degree of degradation of waterlogged wood from different seabed environments.

The preservation status should be classified in terms of both microbial wood decay processes and the distribution of inorganic contaminants, but not only with regard to the content of sulfur and iron. The sulfur and iron penetration profile from the surface throughout the wood has been analyzed and mapped in wood from several shipwrecks found in the Baltic Sea. The results suggest the accumulation of inorganic elements are strongly related to the successive microbiological decay of the waterlogged wood.

However, the degree of preservation of the wood and the distribution of the accumulated sulfur and iron may differ significantly in waterlogged shipwrecks from different seabed environments. Sometimes they also differ in samples from the same wreck.

From these findings it has been concluded that iron and maybe also other inorganic elements in the wood cores may serve as a biochemical fingerprint of the history of the environment on the seabed wreck site.

Different species of sulfur bacteria have been reported that are present in different depths in wood (10, 11). The species segregate and operate at different depths in wood. The interactivity between all these bacteria is mainly unknown but investigations of microbial communities inside the wood indicate a large number of bacterial species.

Synchrotron-based sulfur X-ray absorption spectroscopy has shown a considerable accumulation of organosulfur, pyrite and iron(II) sulfides in marine archaeological wood preserved in seawater. In the museum, the oxidation of these sulfur compounds in the presence of iron ions may cause a severe acidity in the moist wood. The developments of conservation methods to remove acid and iron have been reviewed (12).

The sulfur accumulation in pinewood has been simulated using fresh samples with sulfate-reducing bacteria. The sulfur content in the wood increased more than 10 times in 2 *y* (13).

The depositional process of how wood is incorporated into the wreck site matrix is a geochemical key factor for the preservation of organic material in waterlogged environments (14, 15).

A systematic model in marine environments has been developed that uses physical, biological and chemical processes for the characterization of the wreck site formation with predictive and measurable parameters (16).

Apart from mechanical erosion by the physical transport of sediment, the decisive environmental parameters, directly or indirectly controlling the microbiological degradation activity as well as the number and diversity of microbial species are: Redox potential, composition of organic matter, water input, depth of burial, moisture, oxygen, pH, temperature, salinity, conductivity, as well as the characteristics of the clay or sediment type, such as content, particle size and surface, which governs the ion exchange capacity.

However, the degradation ultimately depends on the availability of electron acceptors and certain chemical species. The oxidants are consumed in order of decreasing energy production per mole of organic carbon oxidized in the order: oxygen, manganese oxides, nitrate, iron oxides, and sulfate. This fact suggests analysis and mapping of inorganic redox elements in the wood related to decreasing redox levels in addition to sulfur and iron.

Investigations of microbial communities inside the wood have indicated a large number of bacterial species (11,17).

Understanding of the biogeochemical processes within the sediment and the interactions with wood in shipwreck timbers under different conditions can increase the possibilities for long-term preservation of these objects (8).

3.1.3 *Analytical Instrumental Techniques to Study Degradation*

In order to preserve degraded wood and to develop and apply suitable conservation treatments, a chemical and physical characterization of archaeological wood is needed. The state of the art of instrumental analytical tools that are available and in use has been reviewed (18). In addition, the applicability of each examined technique is illustrated and discussed through specific examples from the literature. Common analytical instrumental techniques are summarized in Table 3.1.

Table 3.1 Analytical instrumental techniques (18).

Method	Acronym
Scanning electron microscopy	SEM
Fourier transform infrared spectroscopy	FTIR
Nuclear magnetic resonance	NMR
Direct exposure-mass spectrometry	DE-MS
Pyrolysis mass spectrometry	Py-MS
Pyrolysis gas chromatography mass spectrometry	Py-GC/MS

SEM is an excellent method for observing the morphology of the surface of a wood sample. Computed tomography techniques are able to look inside an object and reconstruct its three-dimensional microstructure in the form of imaging. Raman spectroscopy elucidates the organic materials. It does not require any chemical treatment of the sample.

Fourier transform infrared (FTIR) spectra from wood show a fingerprint region at 800–1700 cm^{-1} and a large OH band in the region of 2900–3400 cm^{-1}. Specific bands can be assigned to the wood components, i.e., cellulose, hemicelluloses, and lignin and some degradation products. FTIR has often been used for the semiquantitative evaluation of the degradation state of archaeological wood.

Fourier transform near infrared spectroscopy is a comparatively new tool with a great potential for nondestructive measurement of organic materials. The energy of infrared light stimulates the vibrations of C–O, O–H, C–H, and N–H bonds and results in overtones and combination bands. This depends on the molecular structure, chemical composition or physical properties of the measured surface (19).

With Fourier transform near infrared spectroscopy the aging process of archaeological wood could be assessed. Examples have been detailed (19). It has been shown that the analysis of the spectroscopic data with multivariate analysis is useful for building prediction models for various wood properties (20). Such models are considered as very useful tools for the rapid assessment of the archaeological wood degradation level or the chemical composition of the wood sample (19).

Nuclear magnetic resonance spectroscopy (NMR) is an important analytical technique for the characterization of organic materials. However, its impact on archaeology has been minimal because of its sensitivity and resolution. Recently, instrumental developments have drastically changed this situation, and now NMR is used with good results in the field of cultural heritage. Unilateral NMR is a portable noninvasive technique, which allows *in-situ* measurements without any sampling. It has recently been used for numerous applications in the field of cultural heritage (21). The magnetic field is applied to one side of the object. Therefore, measurements can be performed directly on large objects, such as wall paintings, monuments, and buildings, fully preserving the integrity and the dimension of the object under investigation.

In analytical pyrolysis the object is thermally degraded at 300–800°C. The pyrolysis products provide analytical information on the original sample. The products are preferably analyzed by gas chromatography (GC)-mass spectroscopy (MS) or only by MS. In summary, the degradation product may form a fingerprint that is characteristic of a particular sample, in terms of both the fragment nature and its distribution (18,22).

Evolved gas analysis-mass spectrometry (EGA-MS) was used to study archaeological wood, in order to investigate its chemical degradation (23). The archaeological wood was from an oak pile from a stilt house found in the Neolithic La Marmotta village (Lake Bracciano, Rome, Italy). The sampling was performed from the external to the internal part of the pile, following the annual growth rings in groups of five. In addition, sound oak wood and isolated wood components (holocellulose and cellulose) were also analyzed. The results were used to highlight differences caused by the degradation process.

It has been demonstrated that EGA-MS provides information on the thermochemistry of archaeological wood along with in-depth compositional data due to the use of MS. The wood degradation can be seen in terms of differences between carbohydrates and lignin content. Also, it was shown that lignin oxidation and depolymerization may take place in the archaeological wood. An increase in the formation of wood pyrolysis products bearing a carbonyl group at the benzylic position and a decrease in the amount of lignin dimers were observed. These were related to oxidation and depolymerization reactions, respectively (23).

In addition, wet chemical methods of wood analysis, developed in the pulp and paper industry, can be used to calculate the concentration of wood components after their isolation and purification. However, these classical methods can determine only the amounts of the main wood components: Cellulose, hemicelluloses, lignin, extractives, and ashes (18).

3.1.4 *Near Infrared Spectroscopic Observation of the Aging Process*

The aging degradation of the fine wood structure of dry-exposed archaeological wood was investigated by FTIR near infrared spectroscopy and using a deuterium exchange method (24). The archaeological wood sample was taken from an old wooden temple in Japan of the late 7th century.

Comparing the analytical results with those of a modern wood sample of the same species, the aging process of archaeological wood was clarified as a change in the state of order on a macromolecular structural level.

It could be concluded from the FTIR near infrared spectroscopy spectra that the amorphous region, and partially semicrystalline region, in cellulose, hemicellulose, and lignin decreased by the aging degradation, whereas the crystalline region in cellulose was not affected by the aging.

The accessibility of the diffusant to effect the H/D-exchange was monitored by an OH-related absorption band. This band characteristically varied with the aging process of the wood samples. A morphological model has been proposed to describe the variation of the fine structure of the microfibrils in the cell wall with aging degradation. The state of microfibrils changed loosely by aging, so that elementary fibrils were arranged loosely under 5 Å, whereas several elementary fibrils in the modern wood were arranged in very close proximity under 3 Å to each other (24).

3.1.5 *X-ray Computed Tomography for Anatomical and Dendrochronological Analysis*

Dendrochronology is an important method for dating and provenancing wood samples (25). When this method is used in the conventional way, it would require a large interference with the object.

X-ray computed tomography is a nondestructive technology for the analysis of the ring widths. At the same time it enables the identification of the wood species identification, which is essential for dating.

First attempts for nondestructive tree-ring dating using X-ray computed tomography were already done in the 1980s (26). However, at this time the resolution of the medical scanner employed for the analysis was not sufficient.

Even today, medical computed tomography scanners fail to deliver adequate resolution for noninvasive dating. the noninvasive analysis of timbers only became feasible with the development of micro-X-ray computed tomography.

A still better method is submicron computed tomography. This technology was tested for two specific cases (25): On wood block-lifted together with the surrounding soil and other objects, and, on wood preserved through mineralization on iron objects.

Both micro- and submicron-computer tomography scanners were used in the course of the experiments. The results showed that it is possible to perform the identification of wood based on reconstructed computed tomography scan images.

Under certain conditions this was possible even in soil blocks. The results are dependent upon the resolution and contrast that can be achieved in a measurement, which is affected by both object size and wood species, apart from the characteristics of the scanner (25).

3.1.6 *Relationship Between Underwater Cultural Heritage Deterioration and Marine Environmental Factors*

The relationship between underwater cultural heritage deterioration and marine environmental factors of the Bucentaure and Fougueux sites has been investigated (27). Two warships were sunk in the course of the Battle of Trafalgar in 1805.

The hydrodynamics and the wave dynamics were conducted in the respective areas. For the sediment characterization, a granulometric analysis was done using the dry sieving method.

Annual temperature variations occur from 14°C in March to 24°C in September. Panels of copper did not show a colonization, but panels of pine and oak wood showed a greater colonization (27).

3.1.7 *Characterizing the State of Preservation of Waterlogged Archaeological Wood*

Nondestructive and destructive methods for determining the physical parameters of waterlogged wood such as the porosity, the water content, the wet and dry bulk densities plus the density of the cell wall material have been presented (28). Considerations in respect to determination of masses, volumes and sampling size are discussed.

The influence of the density increment of the water, sorbed to the cell wall material, is discussed and introduced by a model for the constituents of waterlogged wood, including volumes and masses of the cell wall material, the free water in the pores/cell lumen and the sorbed water in the cell wall.

The density increment is introduced as a correction factor in the equations for determination of the physical parameters. If the equations are used without the correction factor they are similar to the most frequently used methods for determining physical parameters of waterlogged wood. By including the correction factor and thereby incorporating the increased density of the sorbed water in the calculations, the presented equations make it possible to correctly determine values for the physical parameters of waterlogged archaeological wood (28).

3.1.8 Oxygen Consumption by Conserved Archaeological Wood

The rates of oxygen consumption have been measured over an extended period of time of conserved, archaeological wood and four samples of fresh, unconserved wood at 50% relative humidity and room temperature (29). Samples from the Swedish warship *Vasa* and the Danish Skuldelev Viking ships were included.

Most samples were preconditioned to 50% relative humidity in a climate chamber to get a stable weight. The chambers were not temperature controlled. For that reason minor fluctuations of the relative humidity could occur. The samples were collected in gas tight bags, filled with oxygen. The volumes of the sealed bags containing the samples were determined by measuring their buoyancy in water.

Most oxygen consumption rates were close to 1 $\mu mg^{-1}d^{-1}$. It was found that the process persisted at least for several years. The consumption of oxygen effects a change in the chemical composition of the wood, which is, in turn, related to degradation. Thus, it has been demonstrated that despite conservation, waterlogged archaeological wood also continues to degrade under museum conditions (29).

3.2 Materials for Conservation

3.2.1 *Dimensional Stabilization*

There are three main ways to deal with the problem of dimensional stabilization of wood containing excessive humidity in order to prevent its destruction at the moment that the water is eliminated (30).

The first way, is the most often used. It involves that category of methods through which gaps in the interior of the wood are filled with solid materials, which resist deformation and are capable of replacing the water in the wood.

The second way consists of eliminating the water from the wood by using an intermediary liquid, which should preferably be nonpolar and have a superficially low tension. When the replacement liquid is done, a porous structure results without dimensional changes, which accompanies the direct evaporation of the water without causing a collapse.

The third way uses the methods of freezing the water from the wood and the sublimation of it in vacuum, i.e., freeze drying.

Within the framework of these ways, until now, over 25 distinct and differentiated methods, depending on the wooden substances, degradation and the dimensions of the wood and the used substances and technologies, are known (30).

The choice of the most adequate methods depends on a multitude of factors and circumstances. In particular, the reactions of the used substances to humidity, the structure and the extent of the degradation of the wood, the microclimatic conditions of the exposition or depositing of the wood, the aesthetic exigencies and last but not least, the economic viability, all play an important role in the decisive act (30).

3.2.2 *Polymers for Archaeological Wood*

There are monographs concerning the archaeological conservation methods using polymers (31). Also, chemical analytical techniques and archaeological interpretation have been described in detail (32–37).

Numerous polymers have been tested in the past, however, most do not stabilize the wood sufficiently. They do not penetrate far enough, or remain stable without producing toxic fumes. Some common examples are (38): Alum salt, $KAl(SO_4)_2 \times 12H_2O$, which was used for treatment earlier but does not penetrate well and leaves the wood very acidic. Poly(ethylene glycol) (PEG) is widely used as a consolidant today, but this material degrades over the course of time and thus cannot support the objects for a very long time. Also, a melamine-formaldehyde resin (Kauramin®) has been used, and while it is fairly stable, it may also fill the wood and turn it into a block of plastic.

So, new consolidants would be advantageous. The requirements of such consolidants have been assessed and it has been discussed how the material sciences may help for making statements. An important requirement for a stabilizing agent is to leave an airy structure in order to allow a future re-treatment. This might be accomplished by foaming a polymer, or by combining nanoparticles with a polymer spider web network to keep them in place. Such particles may help in stabilizing the pH inside the wood by neutralizing any acid generated inside the treated artefacts.

In addition, special attention has been given to the field of biomimetics. This is the discipline of constructing materials inspired by existing natural designs. It may be possible to construct a frame using bioinspired materials (possibly an artificial lignin mixed with other compounds to optimize strength and flexibility) or through biomineralization (an inorganic skeleton).

Tests on biomimetic cellulose and chitosan have started. Chitosan, cf. Figure 3.1, is made from modified chitin (primarily from shrimp and crabs) and may be dissolved in acidic solutions. Crystalline cellulose is attractive in conservation since the individual particles are resistant to acid and not as hygroscopic as the amorphous part of cellulose.

It has been shown that crystalline cellulose particles usually flocculate when used to treat archaeological wood but that they may be treated with surfactants in order to improve the penetration of archaeological objects (38).

3.2.2.1 Novolac Resins

Phenol formaldehyde resins were investigated as consolidants for alum-treated wood (39). X-ray tomography revealed that it is pos-

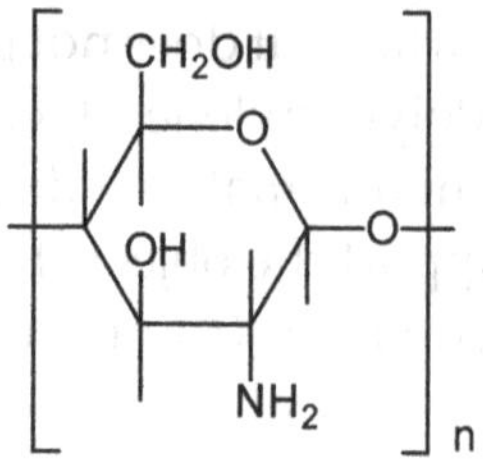

Figure 3.1 Chitosan.

sible to obtain a porous structure inside the wood, ensuring ethical acceptability by allowing a re-treatment.

In order to understand the curing mechanics, the kinetics of the initial condensation reaction were elucidated at room temperature. It was found that a second-order reaction using both phenol and formaldehyde concentrations is the most probable mechanism (39).

3.2.2.2 *Silicone Polymers*

The efficiency of a system containing silanol-terminated poly(dimethylsiloxane), i.e., silanol, a crosslinking agent and a catalyst for the conservation of waterlogged Neolithic archaeological wood, has been evaluated. The conservation procedure requires four basic steps (40):

1. Dehydration of waterlogged wood with acetone,
2. Exchange of acetone with turpentine,
3. Silanol impregnation, and
4. Curing.

The final product after the curing procedure is a three-dimensional polysiloxane network, which in fact constitutes the conservation material. To optimize the procedure, the effect of artefact size, the particular dehydration agent used, and conditions of curing on the antishrink efficiency have been studied. Treated pine and oak waterlogged archaeological wood samples preserved their identity perfectly. No collapsing or shrinking was observed. The color, weight and haptic behavior were those of natural wood (40).

3.2.2.3 Supramolecular Polymer Networks

Three major issues in the conservation of wood are the structural instability on drying, biological degradation, and chemical degradation on account of Fe^{3+} catalyzed production of sulfuric and oxalic acid in waterlogged timbers. It has been stated that before this publication no conservation treatment existed that effectively addresses all three of these issues simultaneously (41).

A conservation treatment has been presented that is based on a supramolecular polymer network constructed from natural polymers with dynamic crosslinking formed by a combination of both host-guest complexation and a strong siderophore pendant from a polymer backbone.

Through a chemical functionalization of chitosan and guar and the addition of a macrocyclic host molecule, i.e., cucurbit[8]uril, the new supramolecular polymer network was developed for the conservation of waterlogged wooden artifacts that simultaneously addresses the three major issues facing conservators.

Cucurbiturils are macrocyclic molecules made of glycoluril $C_4H_2N_4O_2$ monomers linked by methylene bridges –CH2– (42). The oxygen atoms are located along the edges of the band and are tilted inwards, forming a partly enclosed cavity. Glycoluril is shown in Figure 3.2.

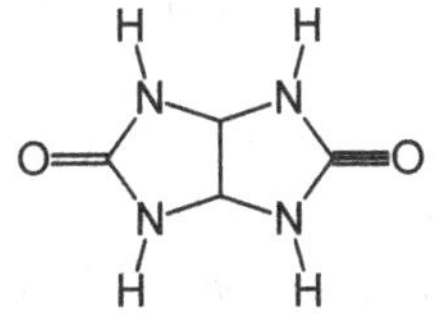

Figure 3.2 Glycoluril.

The functionalization of chitosan was done with both naphthol and catechol. The use of a boronic acid functionalized viologen derivative enabled the attachment to guar through a boronic acid-diol dynamic covalent interaction. Viologens are toxic bipyridinium derivatives of 4,4′-bipyridyl. Coupled with the natural antibacterial properties of chitosan, viologen derivative significantly enhanced the biocidal activity of the entire network. 4,4′-Bipyridyl is shown in Figure 3.3.

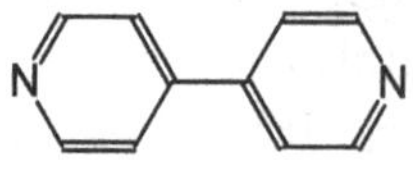

Figure 3.3 4,4′-Bipyridyl.

The supramolecular consolidant has the ability to chelate and trap the iron ions while enhancing the structural stability. The incorporation of antibacterial moieties through a dynamic covalent linkage into the network effects an improved biological resistance in the material.

Exploiting an environmentally compatible natural material with completely reversible chemistries is a safer, greener alternative to current strategies and may extend the lifetime of many culturally relevant waterlogged artifacts around the world (41).

3.2.3 *Nanotechnologies for the Restoration of Archaeological Wood*

3.2.3.1 *Alum-Treated Archaeological Wood*

Alum was widely used during the early 1900s as a treatment for archaeological wood to prevent its shrinkage and to impart strength. In the 1990s, the alarming condition of the wooden objects from the Oseberg burial mound was observed.

Alum solutions which have not been heated exhibit pH levels of 3.5–4. Solutions which have been heated to the temperature used in the alum treatment of 90°C have a pH of about 2 after cooling (43). Thus the wood becomes highly acidic and brittle with little remaining structural integrity.

Initial investigations showed that the previous alum treatment initiated a slow but ongoing deterioration process, attacking the wood for over 100 *y*. The artefacts became highly acidic and had significantly reduced mechanical strength (44).

The use of nonaqueous alkaline nanoparticle dispersions has shown successful results for the protection of cellulose-based materials.

Calcium hydroxide nanoparticles were synthesized in an autoclave system working at high temperature and pressure via a solvothermal process in a two-step reaction. In the first step of the synthesis, calcium metal is oxidized by a short-chain alcohol, such as ethanol, leading to the production of the corresponding alkoxide. The subsequent hydrolysis of this compound generates a colloidal calcium hydroxide, which is dispersed in the solvent for the application. Using this synthetic procedure, stable and highly concentrated calcium hydroxide nanoparticle dispersions can be obtained (45).

Alum-treated archaeological wood samples from Oseberg, with a $pH = 2$, have been treated with alkaline nanoparticle dispersions, and the effects of the treatment have been evaluated by thermal analysis, infrared spectroscopy, and X-ray microtomography (44).

The analysis of the FTIR spectra of alum-treated wood samples clearly shows that alum and wood presented overlapping bands in the region of 1700–1500 and 1200–800 cm^{-1}. The high level of degradation of wood is further confirmed by the lack of the characteristic peaks from cellulose and hemicelluloses, in particular, the signals at 898 cm^{-1}, 1158 cm^{-1}, 1361 cm^{-1}, and 1738 cm^{-1}.

The spectra from the Oseberg sample treated with nanoparticles show the presence of a band at 659 cm^{-1}, which can be attributed to the antisymmetric bending vibration mode of SO_4^{2-}. This obviously results from the neutralization reaction of sulfuric acid with the alkaline compound.

The application of the alkaline nanoparticles dispersed in alcohol on some selected archaeological wooden finds has produced a pH increase of 2–3 (44).

3.2.3.2 *Strontium Carbonate Nanoparticles*

Sulfur compounds, incorporated into the wood prior to its excavation, can lead to acid formation when the samples are exposed to oxygen, and also in the presence of iron ions (46).

Strontium carbonate nanoparticles have been shown to reduce the production of acid formation by their reaction with inorganic sulfur-containing compounds.

The feasibility of using this treatment has been demonstrated on small samples where consolidating treatments have already been performed. It is found that PEG 200 does not prevent the reactivity of the nanoparticles with the sulfur compounds present in the samples.

A surface brushing application method was found to be successful while retaining the visual integrity. Further, it was found that the technique results in the leaching of iron from the surface layers, thus preventing future buildup of acid, catalyzed by iron compounds (46).

3.2.4 *Enzymes for Cleaning*

The effect of hydrolyzing enzymes on the anatomical structure and chemical composition of archaeological wood samples was studied. Pine and beech wood samples, which were taken from anonymous mashrabia, have been cleaned with three types of fungal enzymes. These enzymes were (47):

1. Protease enzyme from *Aspergillus oryzae*. One unit is the amount of enzyme which hydrolyzes 1 $\mu molmin^{-1}$ of *l*-Leucine-*p*-nitroanilide per minute.
2. Lipase enzyme from *Candida rugosa*. One unit will liberate 1 $\mu molmin^{-1}$ oleic acid per minute at pH 8.0 and 40°C.
3. α-Amylase enzyme from *A. oryzae*. One unit will liberate 1 $\mu molmin^{-1}$ maltose per minute at pH 6.0 and 25°C.

Then SEM and FTIR were done in order to assess structural or chemical differences that could be detected between the untreated and the treated wood.

The FTIR indicated no dramatic changes in functional groups on the wood surface before and after enzymatic cleaning. However, the SEM measurements showed that protease and lipase enzymes may give good results in cleaning the wood surface.

However, the enzyme residues remaining on the cell walls of archaeological wood are still another issue (47).

3.2.5 *Chitosan Treatment*

Due to problems with most of the existing conservation treatments, new materials for the stabilization of waterlogged archaeological wood are highly desirable (48). The ideal consolidant should be environmentally friendly and safe to work with for conservators, while strengthening the objects without ruining the visual impression of the wood or filling it in ways that prevent re-treatment in the future.

Chitosan was tested since it is a multifunctional material that not only offers support but also hinders catalytic activity by potentially chelating reactive metal ions. Three solutions of 2% chitosan in 0.1 M acetic acid were chosen as a consolidant for samples from a 1150-year-old piece of waterlogged Viking Age wood from Slagen Prestegård. The penetration of chitosan was determined by freeze drying the samples and analyzing them using high-performance liquid chromatography (HPLC).

Depolymerizing the chitosan only improved uptake slightly. The chitosan left an open structure in the waterlogged archaeological wood. It offered some strength to severely degraded waterlogged wood, but further tests are needed to evaluate stability, shrinkage, and fixation (48).

3.2.6 *Acetone-Carried Consolidants*

Acetone-carried consolidants for waterlogged archaeological wood were tested in order to evaluate these treatments that may save time and energy. Colophony (rosin), two esterified colophonies (Rosin 100® and Rosin 459®), a mixture of colophony with PEG 3400 and a a vinyl acetate-vinyl versate copolymer (Vinavil 8020S®) were tested. Several issues were taken into consideration (49):

- Setting a treatment temperature closer to room temperature, in order to decrease the acetone evaporation and the energy consumption,
- Increasing the plasticity of colophony using a mixture with PEG 3400;
- Evaluating some chemically modified types of colophony commercially available, which possess excellent stability to thermal and oxidative degradation, and
- Evaluating a chemically different type of product, i.e., a vinyl copolymer.

The treatments were carried out at temperatures of 20°C and 35°C on waterlogged maritime pine, elm, oak and beech. The materials were collected from the archaeological site of the ancient ships of

Pisa (Tuscany, Italy) and were dated back to the 7th century BC to the 2nd century AD. The samples were impregnated by immersion in acetone solutions with the polymers mentioned above. Before immersion, the wood was desalinized by washing with deionized water and afterwards the water was replaced by acetone.

To evaluate the processes, the equilibrium moisture content and the dimensional stability of treated wood samples at different relative humidity, and retention of impregnating products were measured. In addition, a macroscopic and a microscopic examination were run to assess the shape and appearance of treated wood and the way of depositing of the consolidants (49).

The results showed that natural and modified colophony treatments gave the most satisfactory results, both in the maintenance of shape and dimensions of samples and also stabilization with respect to variations of the relative humidity. Moreover, the equilibrium moisture contents of samples treated with Rosin 100 and Rosin 459 were strongly reduced in comparison to the other consolidants and to untreated archaeological wood. This finding was explained by the high retention values of those products that occluded most of the porosity, including the microporosity of cell walls (49).

3.2.7 *Natural Polymers as Alternative Consolidants*

Recently ther has been increased interest in examining alternative polymers for the conservation of archaeological artefacts, particularly waterlogged timbers; thus providing better, renewable, greener alternatives to PEG (50).

The degradation of PEG consolidants in the timbers of the 16th century warship the *Mary Rose* has been examined and the rheological and thermal properties of PEG have been compared with its monomethyl and dimethyl ethers and several polysaccharide consolidants, e.g., chitosan, guar, and 2-hydroxyethyl cellulose, in order to evaluate their potential as alternative consolidants for the conservation of waterlogged wooden artifacts. In addition, the effect of the polymers on the archaeological wood was characterized by thermogravimetry and solid-state NMR spectroscopy.

The results suggest that the future of conservation technologies points to polysaccharide consolidant materials, which show an enhanced compatibility with wooden artifacts with no detectable side effects. In addition, these materials are also cheap, with extremely low toxicity, and are renewable and sustainable (50).

3.3 Degradation

3.3.1 *Chemical Changes of Wood by Conservation and Degradation*

The conservation of waterlogged archaeological wooden artefacts from the Nydam Bog in the southern part of Denmark was investigated by Raman spectroscopy. Two different conservation methods were used (51):

1. The cellosolve/petroleum method, and
2. The PEG freeze drying method.

Conservation with cellosolve/petroleum at both room temperature and elevated temperature was used to preserve archaeological arrows made from pinewood and ash wood. Cellosolve is also known as 2-ethoxyethanol.

The double bonds in lignin were affected by the warm treatment. Holocellulose in the archaeological wooden arrows had decomposed completely. Holocellulose is a mixture of cellulose and hemicellulose. Hemicellulose, also known as polyose, is a heteropolymer from several carbohydrates, e.g., xylose, mannose, glucose, and galactose.

A band around 180 cm^{-1} in the R(v)-representation of the low-wavenumber Raman spectrum was used to monitor the presence of water in PEG 2000. The lignin/PEG 2000 relative band intensities were used to make a quantitative estimate of the wood to PEG 2000 ratio in a PEG 2000 impregnated piece of hardwood. A decreasing degradation of holocellulose was observed from the surface to the center of an oak piece from the Kolding Cog, built around 1200 AD. A broad background was observed in the spectra recorded close to the surface of the plank (51).

A similar, very intense, broad background was seen in untreated collapsed ash wood from the Nydam Bog. This broad background, most probably arising from fluorescence, was too intense to allow a Raman spectrum to be seen.

In these cases, ATR-FTIR spectroscopy is a valuable complementary tool for the characterization of waterlogged archaeological wood. It was confirmed that holocellulose degrades more rapidly than lignin (51).

3.3.2 *Microbial Degradation of Waterlogged Archaeological Wood*

The knowledge of degradation processes is essential for the development of conservation methods and *in-situ* preservation of wooden cultural heritage (52,53).

Waterlogged archaeological wood is degraded very slowly in comparison to wood that is located above ground. The special environmental conditions below the ground result in a prolonged decay process that is due to the extremely low oxygen concentration, which only allows a bacterial degradation of wood. Erosion bacteria types have been described. On the other hand, soft rot fungi are microbes that often are found active in a more oxygenated aquatic environment (53).

3.3.3 *Fungi*

Fungi play a considerable role in the deterioration and degradation of cultural heritage due to their enormous enzymatic activity (54). A total of 112 fungal isolates were identificated from selected archaeological wooden objects located at different areas. These areas were the Islamic Art Museum, the storage area of the Solar Boat of King Cheops, the excavation of Saqqara, and the Grand Egyptian Museum.

Aspergillus spp. were predominant in all investigated samples. Thirty seven fungal isolates were screened for cellulose, pectinase and ligninase activity. *Aspergillus brasiliensis* Varga, Frisvad & Janos and *Penicillium duclauxii* Delacroix exhibited high cellulolytic activity, while *Aspergillus amstelodami* (Mangin) Thom & Cruch and *Aspergillus parasiticus* Speare have high pectinolytic activity. Seven fungal species showed ligninolytic potential activity based on their ability to oxidize dyes (54).

Wet organic archaeological materials extracted from seawater may suffer damage as a result of degradation influenced by microorganisms. One of the most common phenomena is indirectly induced by sulfate-reducing bacteria (SRB). Due to their metabolic activity in anoxic conditions, SRB generate hydrogen sulfide from sulfate ions present in seawater. When steel items are in contact with organic matter in the presence of sulfides, corrosion of the metal leads to the precipitation of Fe(II) sulfides. These phases are responsible for dramatic post-excavation damage; their oxidation during storage or exhibition in museums leads to the formation of voluminous crystals, which may cause cracking and crumbling, and lead to the production of sulfuric acid. In order to characterize Fe(II) sulphides and their by-products, 13 waterlogged samples were analyzed by environmental scanning electron microscopy, micro-Raman spectroscopy, and X-ray diffraction. Experiments were performed on untreated wood fragments, on a fragment of rope, and on mineral concretions scratched from the surface of wood remains, all extracted from different shipwrecks. Mackinawite was detected inside the fragments and between the fibers of the rope. Greigite was detected in scattered locations. Pyrite and sulfated phases, like gypsum and iron sulfates, were identified at the surface of the wood fragments and in the mineral concretions.

3.3.4 *Degradation by Microorganisms*

Wet organic archaeological materials extracted from seawater may be degraded by microorganisms (55).

The degradation is most commonly induced by sulfate-reducing bacteria. Due to their metabolic activity in anoxic conditions, sulfate-reducing bacteria generate hydrogen sulfide from the sulfate ions that are commonly present in seawater.

When steel or iron materials are in contact with organic matter in the presence of sulfides, the corrosion of the metal leads to the precipitation of Fe^{2+} sulfides.

These materials are responsible for a dramatic post-excavation damage (55). Their oxidation during storage or exhibition in museums leads to the formation of voluminous crystals, which may cause cracking and crumbling. Also, sulfuric acid is formed. In order to characterize Fe^{2+} sulfides and their by-products, 13 waterlogged samples were analyzed by environmental scanning electron microscopy, micro-Raman spectroscopy, and X-ray diffraction.

The experiments were performed on untreated wood fragments, on a fragment of rope, and on mineral concretions scratched from the surface of wood remains, all extracted from different shipwrecks.

Mackinawite was detected inside the fragments and between the fibers of the rope. Furthermore, greigite was detected in scattered locations. Pyrite and sulfated phases, like gypsum and iron sulfates, were identified at the surface of the wood fragments and in the mineral concretions (55).

Mackinawite is an iron nickel sulfide mineral with the formula $(Fe, Ni)_1 + xS$, where $x = 0–0.11$ (56). Mackinawite occurs among others in reducing environments, such as marine sediments, as a result of the metabolism of iron- and sulfate-reducing bacteria. Depending on the redox conditions, mackinawite can form more stable phases such as greigite and ultimately pyrite (57).

Greigite is an iron sulfide mineral with formula Fe_3S_4. It is the sulfur equivalent of the iron oxide Fe_3O_4, i.e., magnetite. It was first described in 1964 for an occurrence in San Bernardino County, California, and named after the mineralogist and physical chemist Joseph W. Greig (58).

3.3.5 *Degradation of Archaeological Wood Under Freezing and Thawing Conditions*

Permafrost can ensure an excellent conservation of organic archaeological materials (59). However, nowadays, the temperature is rising and the permafrost region is currently retreating (60), which is a particular threat to Arctic sites, where the preservation of organic material relies on generally low or freezing temperatures. Thus, it is important to know the consequences of permanently frozen archaeological sites.

The degradation of archaeological wood at freezing and thawing temperatures has been studied at the site of Qajaa in West Greenland (59). Environmental monitoring, measurement of oxygen consumption and microscopy of the wood samples have been done.

Permanently frozen wood is still very well preserved after 2–4,000 y. On the other hand, wood samples that thaw every summer show an attack by soft rot and an average density loss of 0.1 gcm^{-3}, corresponding to 25% of the dry mass over the past 27 y.

It has been concluded that future increases in temperature may increase the decay rate significantly: however, the effects on site depend on the local hydrology (59).

3.3.6 *Abiotic Chemical Degradation*

The process of abiotic chemical degradation within a piece of wood has been described using a series of tissues taken at increasing distances from the surface of waterlogged archaeological wood of *Quercus* sp., *Fraxinus excelsior*, *Ulmus campestris*, *Populus alba* and *Picea abies* (61).

Light microscopy and SEM in addition to chemical analysis of the respective tissues, showed that the degradation advanced inwards from the wood surface. Following excessive swelling, the secondary cell wall starts to loosen because of the hydrolysis of the carbohydrates. The cell walls loose their fluorescence and birefringence. Finally, the residual lignin skeleton collapses and only granular debris is left.

Tertiary walls and, in particular, the system of the compound middle lamellae, are resistant to degradation and retain the original dimensions of the tissue as long as it is filled with water (61).

3.3.7 *Degradation of Lignin in Archaeological Waterlogged Wood*

Archaeological wooden objects are generally better conserved in wet environments than in other environments (62). Nevertheless, anaerobic erosion bacteria can slowly degrade waterlogged wood, causing a loss of cellulose and hemicellulose and leading to the formation of water-filled cavities.

In the course of this process, lignin can be also altered. Thus a porous and fragile structure is formed, poor in polysaccharides and mainly composed of residual lignin, which can easily collapse during drying and needs specific consolidation treatments.

Lignin is a complex three-dimensional macromolecule, made up of various different monomeric units, basically derived from coniferyl alcohol in softwoods, and from both coniferyl alcohol and sinapyl alcohol in hardwoods (63,64). These alcohols are shown in Figure 3.4.

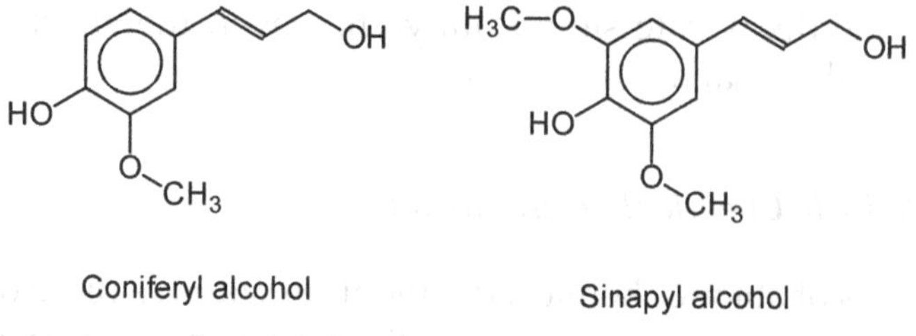

Figure 3.4 Coniferyl alcohol and sinapyl alcohol.

For the purpose of lignin degradation, the chemical characterization of archaeological lignin is of primary importance in the diagnosis and conservation of waterlogged wood samples (62).

Hitherto, knowledge of the lignin degradation processes in historical and archaeological wood has been extremely inadequate. Lignin extracted from archaeological waterlogged wood was examined using both pyrolysis GC MS, NMR spectroscopy and gel permeation chromatography (GPC).

The samples were collected from the site of the Ancient Ships of San Rossore (Pisa, Italy), where since 1998 31 shipwrecks, dating from the 2nd century BC to 5th century AD, have been discovered.

The results, using GPC analysis, highlight the depolymerization of lignin with the cleavage of ether bonds, leading to a higher amount of free phenol units in the lignin from archaeological waterlogged wood, compared to lignin from a reference wood of the same species (62).

3.3.8 *Identification of Bacterial Cultures*

Anaerobic bacteria were isolated from a 1700 y old wooden spear shaft, excavated from an archaeological site that dates from the Iron Age, in the southern part of Jutland, Denmark (65).

The bacteria were cultivated in glucose-supplemented and xylose-supplemented media at 14°C and 20°C. A gene library with 21 clones was constructed by extracting and amplifying 16S rDNA sequences from the individual cultures. One clone was phylogenetically affiliated to the *Spirochaeta* type. Eleven clones affiliated with an unidentified member of the α-proteobacteria were present in all the culture samples. Three clones were affiliated with the β-proteobacteria. Four clones were clustered among the *Geobacteriaceae,* in the β-proteobacteria. A single clone was clustered with Gram-positive bacteria. All the identified bacterial families are commonly found in soil or bog environments. Many of them can use cellulose as their carbon or energy source (65). So, these bacteria may degrade the wood.

3.4 Special Properties

3.4.1 *Wooden Shipwrecks*

Wooden shipwrecks in the marine environment form a large part of the underwater cultural heritage (15). Over the past 50 years several wrecks have been excavated, raised and conserved. In the recent past there has been a trend towards preserving these sites *in-situ,* on the seabed, as opposed to raising them.

The deterioration of wood in the marine environment has been described. Further, the principles of the most commonly used methods for conserving waterlogged archaeological wood have been reviewed. Also, a general approach to tackling the *in-situ* preservation of wooden wreck sites has been presented (15).

3.4.2 *State of Preservation of Waterlogged Archaeological Wood*

The state of preservation of waterlogged archaeological wood samples was evaluated based on the maximum moisture content (MMC), the basic density and the wood substance density (WSD) determined in water and helium (66). The degree of wood degradation was compared under the criteria: The loss of wood substance (LWS) and the loss of wood substance density.

The MMC is calculated as:

$$MMC = 100\frac{m_w - m_d}{m_d} \tag{3.1}$$

Here, m_w is the mass of the waterlogged sample and m_d is the mass of the oven-dried sample.

The LWS is calculated as:

$$LWS = 100\frac{\rho_f - \rho_a}{\rho_f} \tag{3.2}$$

Here, ρ_f is the basic density of fresh wood and ρ_a is the basic density of archaeological wood.

Studies were done on several wood samples differing in species, degree of degradation, age and their place of origin. The physical properties of wood were determined for the material containing mineral compounds and the material without minerals. The properties of the latter, in which the minerals are replaced by water, were calculated from the mass and volume of the wood containing minerals, as well as, the content and the density of the ash sampled after burning the sample.

The study revealed the effect of minerals on the tested parameters and wood degradation indices. A strong relationship between the basic density and MMC was confirmed for both the wood containing minerals and without them; by contrast a substantially weaker correlation between basic density and WSD was observed.

Furthermore, it was found that the assessment of the state of wood preservation conducted on the basis of WSD and loss of wood substance density yielded different results. Also, it seems that both indices of wood degradation might be unreliable. The main drawback of the LWS-based assessment is associated with a wide range of basic densities of fresh wood. In turn, the loss of wood substance density mainly indicates the changing ratio of the carbohydrates and lignin content, but it fails to provide information on the loss of wood substance.

This may hinder the comparison of the wood sampled from different sites and subjected to different decay mechanisms. Nevertheless, the WSD-based assessment of the state of preservation of waterlogged archaeological wood might be a valuable complementary method to the basic density or MMC-based assessment, which is routinely carried out in many conservation centers (66).

3.4.3 *Adsorption and Desorption Mechanism of Water*

The adsorption and desorption mechanism of water was investigated for modern and archaeological wood by near infrared spectroscopy (67). A mixture model of water was used to decompose the near infrared difference spectra into three components (67):

1. Free water molecules,
2. Water molecules with one OH group engaged in hydrogen bonding, and
3. Water molecules with two OH groups engaged in hydrogen bonding.

The data were analyzed using principal component analysis. The variations of each water component with relative humidity could be explained by a model that describes water absorption in three stages.

It was concluded that the aging phenomenon in wood is due to the decrease of adsorption sites on hemicellulose and amorphous cellulose (67).

3.4.3.1 *Hygroscopic Properties of PEG-Treated Archaeological Wood*

An archaeological wood from the rampart of the Poznań stronghold, built in the 10th century, was tested with respect to its hygroscopic properties. This material was obtained during the excavations and conservation works of the earthen and wooden relicts.

The archaeological wood, both untreated and treated with PEG, was characterized by determining the adsorption and desorption isotherms, which were formed by GAB and GDW models.

The GDW model is known as the generalized D'Arcy and Watt model and the GAB is the Guggenheim-Anderson-de Boer model (68). This model can be used to correlate the water activity and moisture content during an osmotic dehydration process (69).

The GAB equation is:

$$X = \frac{CkX_0a_w}{(1 - ka_w)(1 - ka_w + Cka_w)} \tag{3.3}$$

Here, C and k are constants and X_0 is the monolayer moisture content.

The GDW equation is:

$$X = \frac{CkX_0a_w(1 - k(1 - w)a_w)}{(1 + ka_w)(1 - ka_w)} \tag{3.4}$$

The anomalous increase of the equilibrium moisture content for air relative humidity values above 80% was found for the PEG-treated wood. It was documented that the safety of wood exposition strongly depends on the proper maintaining of air parameters. The mechanisms of the reduction of equilibrium moisture content and sorption hysteresis for the PEG-treated wood were explained in detail. It is not recommended to predict equilibrium moisture content values with the Hailwood-Horrobin model fitted to the sorption data collected by the US Forest Products Laboratory, as it did not account for sorption hysteresis, decay of archaeological wood and the influence of the PEG treatment (70).

3.4.4 PEG-Impregnated Waterlogged Archaeological Wood

To prevent the deformation and cracking of waterlogged archaeological wood, PEG is commonly applied as a bulk impregnation agent (71). PEG maintains the wood in a swollen state during drying. However, the swelling of wood can reduce its mechanical properties.

The cellular structure of oak and cell wall swelling was characterized by SEM of transverse cross sections, and the microfibril angle of oak fibers was determined by wide-angle X-ray scattering.

Samples of recent European oak (*Quercus robur* L.) impregnated with PEG with a molecular weight of 600 *Dalton* were tested in axial tension and radial compression. Mechanical tests showed that axial tensile modulus and strength were only slightly affected by PEG, whereas the radial compressive modulus and the yield strength were reduced by up to 50%.

This behavior can be explained by the microstructure and deformation mechanisms of the material (71). Microfibril angles in tensile test samples were close to zero. This implies tensile loading of cellulose microfibrils within the fiber cell walls without almost any shear in the adjacent amorphous matrix.

These results can help to separate the impact of PEG on the mechanical properties from that of chemical degradation in archaeological artifacts, which display only a small biological degradation (71).

3.4.5 *Patterns in Tree Rings*

A method has been described as to how tree-ring data from subfossil and archaeological wood can be interpreted in palaeoecological terms. An attempt was done to provide a general theoretical framework for structuring and interpreting non-recent data, where the ecological context is usually missing.

Here, the most important tools are dendrotypology and the systematic analysis of tree-ring patterns done in four levels. These levels are (72):

1. Ring-anatomical patterns,
2. Impact patterns in ring width series at the single-tree level,
3. Stand-dynamic patterns, i.e., patterns in ring width series on a stand-wide level, and
4. Complex patterns that are made up of consistent combinations of the other three pattern levels with each other or additional information.

The distinction of these patterns allows the disentangling of different sources of influence on tree growth, which can also potentially be applied to archaeological wood.

Several types of reactions to different categories of impact are defined. On a stand-wide scale, the phases of stand development can be reconstructed by identifying characteristic combinations of events. Thus, the range of possible interpretations can be narrowed down to a few possibilities, and sometimes only one.

While the general concept is thought to be applicable to a wider range of regions and ecological settings, the interest is focused on the gap-phase dominated lowland forests and anthropogenic stands derived from such, which form the focus of the Central European dendroarchaeological research (72).

3.4.6 *Physical and Mechanical Properties of Archaeological Wood*

The effects of aging and deterioration on the density, sorption characteristics, shrinkage and swelling, and mechanical properties of both dry and waterlogged archaeological wood have been reviewed (73).

3.4.7 *Demethylation of Syringyl Moieties in Archaeological Wood*

Archaeological oak wood samples from *Quercus* sp., ranging from the 16th C. AD to 6000 BP, were studied using flash pyrolysis GC/MS (74). In this way, an insight into the angiosperm lignin degradation was obtained. The pyrolysates revealed the occurrence of a number of 3-methoxy-1,2-benzenediol derivatives, methoxycatechols, directly related to 2,6-dimethoxyphenol, i.e., syringyl, moieties. These are the characteristic building blocks of angiosperm lignin. 2,6-Dimethoxyphenol is shown in Figure 3.5.

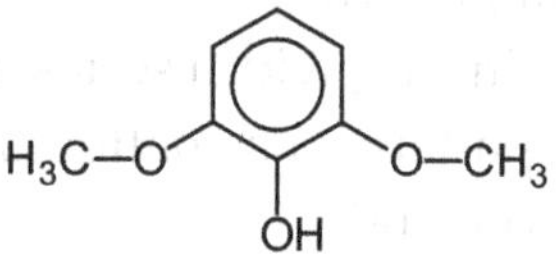

Figure 3.5 2,6-Dimethoxyphenol.

The finding of these characteristic pyrolysis products in well-preserved archaeological wood provides unequivocal evidence that the demethylation of the syringyl units occurs very early in the course of the degradation of wood.

It seems that the absence of abundant 3-methoxy-1,2-benzenediols in the degrading plant materials containing angiosperm lignin relates to the lability of these newly formed moieties.

3.4.8 *Decay Prevention Using Gamma Irradiation*

Gamma irradiation gas has been evaluated for decay prevention treatment of waterlogged archaeological wood (75). A dose of 15 *kGy* was found to be sufficient to inactivate a large number of wood biodeteriogens, including fungi, bacteria and invertebrates, at various stages of development. For timbers excavated from polluted sites, a dose of 25 *kGy* has been suggested to inactivate human pathogens.

The dose spread required for such treatments are 1.33:1 and 1.2:1, respectively, in timbers up to 150 *mm* thickness and a density not exceeding 1590 kgm^{-3}. No adverse effects on the physical properties of slightly or heavily degraded waterlogged archaeological wood could be detected at doses of up to 100 *kGy*. This is the maximum recommended single or cumulative lifetime dose for any timber. It has been found that gamma irradiation offers great efficacy (75).

In another study, gamma irradiation was applied to decontaminate a Korean cultural artifact, a wooden cashbox stored in a local museum (76). Fungi isolated from the wooden cashbox were identified by 18S rDNA sequencing methods.

The isolated fungi exhibited a high similarity to *Aspergillus niger*, *Penicillium verruculosum*, and *Trichoderma viride*. Each strain was tested for its sensitivity to gamma irradiation, and could be inactivated by the irradiation at a dose of 5 *kGy*. Therefore, the wooden cashbox was gamma-irradiated at this dose of 5 *kGy*, and consequently decontaminated.

Two months after the irradiation, when the wooden cashbox was retested to detect biological contamination, no fungi were found. Therefore, these results suggest that gamma irradiation at a low dose of 5 *kGy* can be applied for successful decontamination of wooden artifacts (76).

References

1. R.M. Rowell and R.J. Barbour, eds. *Archaeological Wood: Properties, Chemistry, and Preservation*, number 225 in Advances in Chemistry Series, Washington, DC, 1990. National Meeting of the American Chemical Society, Los Angeles, California, American Chemical Society.
2. V. Dorge, *Painted Wood: History & Conservation*, Getty Conservation Institute, Los Angeles, 1998.
3. B. Rodgers, *The Archaeologist's Manual for Conservation: A Guide to Non-toxic, Minimal Intervention Artifact Stabilization*, Kluwer Academic/Plenum Publishers, New York, 2004.
4. J. Łucejko, *Waterlogged Archaeological Wood Chemical Study of Wood Degradation and Evaluation of Consolidation Treatments*, VDM Verlag Dr. Müller, Saarbrücken, 2010.
5. A. Celant and G. Coccolini, *Plant microtechniques and protocols* in T.E.C. Yeung, C. Stasolla, J.M. Sumner, and Q.B. Huang, eds., *Archaeological Wood Preparation*, chapter 26, pp. 487–493. Springer International Publishing, Cham, 2015.
6. N. Meeks, *Historical Technology, Materials and Conservation: SEM and Microanalysis*, Archetype, London, 2012.
7. M. Hamed, S. Abdelkader, M.F. Ali, N. Elhadidi, and N. Mohamed, *International Journal of Conservation Science*, Vol. 4, 2013.

8. Y. Fors, Biochemical fingerprints in marine-archaeological wood–an overview of a research project, in K. Piotrowska and P. Konleczny, eds., *Condition 2015 Conservation and Digitalization*, pp. 79–82, Gdansk, 2015. Natonal Maritime Museum.
9. P.N. Froelich, G.P. Klinkhammer, M.L. Bender, N.A. Luedtke, G.R. Heath, D. Cullen, P. Dauphin, D. Hammond, B. Hartman, and V. Maynard, *Geochimica et Cosmochimica Acta*, Vol. 43, p. 1075, July 1979.
10. T. Yamashita, R. Yamamoto-Ikemoto, and J. Zhu, *Bioresource Technology*, Vol. 102, p. 2235, February 2011.
11. S. Bessette, S.K. Fagervold, C. Romano, D. Martin, N. Le Bris, and P.E. Galand, *Journal of Marine Science and Technology*, Vol. 22, p. 60, 2014.
12. Y. Fors and M. Sandström, *Chem. Soc. Rev.*, Vol. 35, p. 399, 2006.
13. Y. Fors, T. Nilsson, E.D. Risberg, M. Sandström, and P. Torssander, *International Biodeterioration & Biodegradation*, Vol. 62, p. 336, December 2008.
14. C. Caple, *International Biodeterioration & Biodegradation*, Vol. 34, p. 61, January 1994.
15. D. Gregory, P. Jensen, and K. Strætkvern, *Journal of Cultural Heritage*, Vol. 13, p. S139, 2012. Wood Science for Conservation.
16. I.A.K. Ward, P. Larcombe, and P. Veth, *Journal of Archaeological Science*, Vol. 26, p. 561, May 1999.
17. E.T. Landy, J.I. Mitchell, S. Hotchkiss, and R.A. Eaton, *International Biodeterioration & Biodegradation*, Vol. 61, p. 106, January 2008.
18. J.J. Łucejko, F. Modugno, E. Ribechini, D. Tamburini, and M.P. Colombini, *Applied Spectroscopy Reviews*, Vol. 50, p. 584, May 2015.
19. A. Sandak, J. Sandak, M. Zborowska, and W. Prądzyński, *Journal of Archaeological Science*, Vol. 37, p. 2093, September 2010.
20. J. Rodrigues, A. Alves, H. Pereira, D. da Silva Perez, G. Chantre, and M. Schwanninger, *Holzforschung*, Vol. 60, January 2006.
21. N. Proietti, D. Capitani, and V. Di Tullio, *Sensors*, Vol. 14, p. 6977, April 2014.
22. J.J. Łucejko, F. Modugno, E. Ribechini, and J.C. del Río, *Analytica Chimica Acta*, Vol. 654, p. 26, November 2009.
23. D. Tamburini, J.J. Łucejko, E. Ribechini, and M.P. Colombini, *J. Mass Spectrom.*, Vol. 50, p. 1103, August 2015.
24. S. Tsuchikawa, H. Yonenobu, and H.W. Siesler, *Analyst*, Vol. 130, p. 379, 2005.
25. J. Stelzner and S. Million, *Journal of Archaeological Science*, Vol. 55, p. 188, March 2015.
26. M. Onoe, J.W. Tsao, H. Yamada, H. Nakamura, J. Kogure, H. Kawamura, and M. Yoshimatsu, *Nuclear Instruments and Methods in Physics Research*, Vol. 221, p. 213, March 1984.

27. T. Fernández-Montblanc, M. Bethencourt, A. Izquierdo, and M.M. González-Duarte, Establishing the relationship between underwater cultural heritage deterioration and marine environmental factors. A comparative analysis of the Bucentaure and Fougueux sites, in M. Candelera, ed., *Science, Technology and Cultural Heritage*, pp. 53–58, Leiden, The Netherlands, 2014. Proceedings of the Second International Congress on Science and Technology for the Conservation of Cultural Heritage, CRC Press.
28. P. Jensen and D.J. Gregory, *Journal of Archaeological Science*, Vol. 33, p. 551, April 2006.
29. M.N. Mortensen and H. Matthiesen, *Analytical and Bioanalytical Chemistry*, Vol. 405, p. 6373, May 2013.
30. D.M. Creangă, *European Journal of Science and Theology*, Vol. 5, p. 57, 2009.
31. C. Smith, *Archaeological Conservation Using Polymers: Practical Applications for Organic Artifact Stabilization*, Texas A&M University Press, College Station, 2003.
32. A.M. Pollard, *Archaeological Chemistry*, Royal Society of Chemistry, Cambridge, 1996.
33. M. Glascock, R.J. Speakman, and R.S. Popelka-Filcoff, eds., *Archaeological Chemistry: Analytical Techniques and Archaeological Interpretation*, American Chemical Society, Distributed by Oxford University Press, Oxford, New York, 2007.
34. Z. Goffer, *Archaeological Chemistry*, Vol. 170 of *Chemical Analysis*, Wiley-Interscience, Hoboken, N.J, 2nd edition, 2007.
35. T. Price and J.H. Burton, *An Introduction to Archaeological Chemistry*, Springer, New York, 2011.
36. M. Malainey, *A Consumer's Guide to Archaeological Science: Analytical Techniques*, Springer, New York, 2011.
37. A.M. Pollard, *Archaeological Chemistry*, Royal Society of Chemistry, Cambridge, 2016.
38. M. Christensen, H. Kutzke, and F.K. Hansen, *Journal of Cultural Heritage*, Vol. 13, p. S183, September 2012.
39. M. Christensen, F.K. Hansen, and H. Kutzke, *Archaeometry*, Vol. 57, p. 536, April 2014.
40. P.K. Kavvouras, C. Kostarelou, A. Zisi, M. Petrou, and G. Moraitou, *Studies in Conservation*, Vol. 54, p. 65, 2009.
41. Z. Walsh, E.-R. Janeček, J.T. Hodgkinson, J. Sedlmair, A. Koutsioubas, D.R. Spring, M. Welch, C.J. Hirschmugl, C. Toprakcioglu, and J.R. Nitschke, *Proc. Natl. Acad. Sci., USA*, Vol. 111, p. 17743, November 2014.
42. Wikipedia, Cucurbituril — wikipedia, the free encyclopedia, 2015. [Online; accessed 29-April-2016].

43. S. Braovac and H. Kutzke, *Journal of Cultural Heritage*, Vol. 13, p. S203, September 2012.
44. F. Andriulo, S. Braovac, H. Kutzke, R. Giorgi, and P. Baglioni, *Applied Physics A*, Vol. 122, p. 1, 2016.
45. G. Poggi, N. Toccafondi, L.N. Melita, J.C. Knowles, L. Bozec, R. Giorgi, and P. Baglioni, *Applied Physics A*, Vol. 114, p. 685, 2013.
46. E.J. Schofield, R. Sarangi, A. Mehta, A.M. Jones, A. Smith, J.F.W. Mosselmans, and A.V. Chadwick, *Journal of Cultural Heritage*, Vol. 18, p. 306, 2016.
47. S.A.E.-K.M. Hamed, *Journal of Archaeological Science*, Vol. 39, p. 2515, July 2012.
48. M. Christensen, E. Larnøy, H. Kutzke, and F.K. Hansen, *Journal of the American Institute for Conservation*, Vol. 54, p. 3, 2015.
49. G. Giachi, C. Capretti, I.D. Donato, N. Macchioni, and B. Pizzo, *Journal of Archaeological Science*, Vol. 38, p. 2957, November 2011.
50. Z. Walsh, E.-R. Janeček, M. Jones, and O.A. Scherman, *Studies in Conservation*, Vol. 62, p. 1, 2016.
51. M. Christensen, M. Frosch, P. Jensen, U. Schnell, Y. Shashoua, and O.F. Nielsen, *Journal of Raman Spectroscopy*, Vol. 37, p. 1171, 2006.
52. N.B. Pedersen, C. Björdal, P. Jensen, and C. Felby, *Bacterial degradation of archaeological wood in anoxic waterlogged environments* in S.E. Harding, ed., *Stability of Complex Carbohydrate Structures. Biofuels, Foods, Vaccines and Shipwrecks*, chapter 13, pp. 160–187. RSC Publishing, Cambridge, 2013.
53. C. Gjelstrup Björdal, *Journal of Cultural Heritage*, Vol. 13, p. S118, September 2012.
54. M.E.-S. Osman, A.A.E.-N. El-Shaphy, D.A. Meligy, and M.M. Ayid, *International Journal of Conservation Science*, Vol. 5, p. 295, 2014.
55. C. Rémazeilles, K. Tran, E. Guilminot, E. Conforto, and P. Refait, *Studies in Conservation*, Vol. 58, p. 297, October 2013.
56. Wikipedia, Mackinawite — wikipedia, the free encyclopedia, 2016. [Online; accessed 29-April-2016].
57. D. Csákberényi-Malasics, J.D. Rodriguez-Blanco, V.K. Kis, A. Rečnik, L.G. Benning, and M. Pósfai, *Chemical Geology*, Vol. 294-295, p. 249, February 2012.
58. Wikipedia, Greigite — wikipedia, the free encyclopedia, 2016. [Online; accessed 29-April-2016].
59. H. Matthiesen, J.B. Jensen, D. Gregory, J. Hollesen, and B. Elberling, *Archaeometry*, Vol. 56, p. 479, 2014.
60. B. Metz, *Climate change 2007 Mitigation of climate change: Contribution of Working Group III to the Fourth Assessment Report of the Intergovernmental Panel on Climate Change*, Cambridge University Press, Cambridge, New York, 2007.

61. P. Hoffmann and M.A. Jones, *Structure and degradation process for waterlogged archaeological wood* in R.M. Rowell and R.J. Barbour, eds., *Archaeological Wood: Properties, Chemistry, and Preservation*, number 225 in Advances in Chemistry Series, pp. 35–65. American Chemical Society, 1990.
62. M.P. Colombini, J.J. Lucejko, F. Modugno, M. Orlandi, E.-L. Tolppa, and L. Zoia, *Talanta*, Vol. 80, p. 61, November 2009.
63. D. Fengel, *Wood Chemistry, Ultrastructure, Reactions*, Walter de Gruyter, Berlin, New York, 1989.
64. J. Ralph, K. Lundquist, G. Brunow, F. Lu, H. Kim, P.F. Schatz, J.M. Marita, R.D. Hatfield, S.A. Ralph, and J.H. Christensen, *Phytochemistry Reviews*, Vol. 3, p. 29, January 2004.
65. A. Christine Helms, A. Camillo Martiny, J. Hofman-Bang, B. K. Ahring, and M. Kilstrup, *International Biodeterioration & Biodegradation*, Vol. 53, p. 79, March 2004.
66. L. Babiński, D. Izdebska-Mucha, and B. Waliszewska, *Journal of Archaeological Science*, Vol. 46, p. 372, June 2014.
67. T. Inagaki, H. Yonenobu, and S. Tsuchikawa, *Appl. Spectrosc.*, Vol. 62, p. 860, Aug 2008.
68. S. Furmaniak, A.P. Terzyk, L. Czepirski, E. Komorowska-Czepirska, J. Szymonska, and P.A. Gauden, *Water sorption of foodstuffs – alternative models* in V. Pletney, ed., *Focus on Food Engineering Research and Developments*, chapter 11, pp. 497–515. Nova Science Publishers, New York, 2007.
69. F. Prothon and L.M. Ahrné, *Journal of Food Engineering*, Vol. 61, p. 467, February 2004.
70. W. Olek, J. Majka, A. Stempin, M. Sikora, and M. Zborowska, *Journal of Cultural Heritage*, Vol. 18, p. 299, March 2016.
71. I. Bjurhager, J. Ljungdahl, L. Wallström, E.K. Gamstedt, and L.A. Berglund, *Holzforschung*, Vol. 64, January 2010.
72. N. Bleicher, *Veget. Hist. Archaeobot.*, Vol. 23, p. 615, June 2013.
73. A.P. Schniewind, *Physical and mechanical properties of archaeological wood* in R.M. Rowell and R.J. Barbour, eds., *Archaeological Wood: Properties, Chemistry, and Preservation*, number 225 in Advances in Chemistry Series, pp. 87–109. American Chemical Society, 1990.
74. P.F. van Bergen, I. Poole, T.M. Ogilvie, C. Caple, and R.P. Evershed, *Rapid Communications in Mass Spectrometry*, Vol. 14, p. 71, 2000.
75. S.B. Pointing, E.B. Jones, and A.M. Jones, *International Biodeterioration & Biodegradation*, Vol. 42, p. 17, July 1998.
76. J.-I. Choi and S. Lim, *Radiation Physics and Chemistry*, Vol. 118, p. 70, January 2016.

4

Fossils

4.1 Monograph

There are monographs concerning the preparation and conservation of animal fossils (1–3).

Also, the recent microscopic, chemical and molecular methods for examining fossil preservation have been reviewed (4).

4.2 Paleontological Skill and the Role of the Fossil Preparator

Some practical and philosophical aspects of fossil preparation have been discussed (5). Despite the widespread interest in paleontology, only a few people know how paleontologists produce knowledge about past life. Fossil preparation and the processes done to make fossils useful for research and exhibition have been studied and interpreted. It has been argued that preparators serve as mediators between nature and researchers.

The preparators judge where fossils need internal support, e.g., an adhesive, and external support, e.g., custom-fit rigid molds, based on their knowledge of physics and weight distribution.

Jacketing a fossil involves applying a paste, traditionally a plaster and burlap mixture, and more recently a resin over a foil-wrapped fossil. The paste is allowed to harden and to create an exact mold.

In August 2006, the fossil skull of *Rhomaleosaurus cramptoni* arrived at the Natural History Museum in London to be prepared by Scott Moore-Fay. This 178 million-year-old marine reptile has a long, distinguished history in the scientific and social worlds as well as the geological one. Moore-Fay has made a skull's jacket of epoxy resin, which is more chemically stable than a similar polyester resin (6).

Upon curing, a a super hard jacket is created, which is expected to be good for more than a hundred years. Apart from the use of modern materials, the jacketing technique has changed only a little over time (5).

4.3 Analysis Methods

4.3.1 *Bone Samples*

In a good state of preservation, bone conserves the entire protein pattern of extracellular bone matrix proteins over thousands of years. The quality of the profiles of matrix proteins isolated from ancient bones, separated by electrophoresis, is as good as those from recent bones. Molecules arising from collagenous proteins, e.g., collagen type I, from the noncollagenous group, e.g., osteonectin, and from the immune system, e.g., immunoglobulin G, could be identified in western blots by specific antibodies (7).

Type I collagen is the major protein in bones (8). The mineral matrix protects collagen from denaturation, thus permitting the recovery of large collagen peptides from fossil bones thousands or millions of years old. Collagen peptides are usually more or less denatured in fossil bones. Diagenetic alteration is the major cause of such a denaturation. Classical extraction methods alter the remaining large collagen peptides by extensive solubilization.

A method has been described that uses shorter collagen solubilization times. The thus obtained collagen yields could be favorably compared with classical methods. In particular, the size of the large peptide fraction is considerably improved. Combined with a particular concentration step, the use of this shorter solubilization technique should be useful for a collagen analysis that needs large peptides, e.g., in palaeoimmunology (8).

4.3.2 *Stable Isotope Analysis*

Stable isotope ratios of carbon and nitrogen in bone have become increasingly important over the past few decades. The majority of such studies have been done in North America, where the use of marine foods and the introduction of maize have been monitored. In Europe, stable isotope ratios have documented the shift from marine fishing and gathering to agriculture in coastal areas and the introduction of millet in parts of eastern and southern Europe. A lot of the past studies have been reviewed (9).

Stable carbon ^{13}C and nitrogen ^{15}N isotope analysis is useful in addressing some fundamental questions in ecology such as reconstructing trophic interactions, habitat connections and climate regime shifts (10).

The temporal scales over which stable isotope analysis can be used to address ecological problems vary, depending on the protein turnover times of the analyzed tissue. Hard, inert tissues, such as teeth, bones and mollusc shells, grow in regular intervals, i.e., daily or annually. So, sequential sampling of these growth layers provides a time series of the isotopic patterns. As a result, stable isotope analysis on these tissues is useful for elucidating the behavior and ecology of animals over the course of time, especially those with cryptic life history stages, such as marine turtles that retain growth layers in their humerus bones (10).

Based on the stable isotope analysis of historical bone, it has been assumed that fresh bone samples must be pretreated with acid prior to stable isotope analysis (11).

Using an elemental analyzer coupled to an isotope ratio mass spectrometer to measure the stable carbon and the nitrogen ratios, the need to acidify a cortical bone powder with 0.25 *M* HCl prior to stable isotope analysis has been tested. The goal was to isolate bone collagen for the determination of ^{13}C and ^{15}N values.

It was found that the acidification of micromilled cortical bone samples from marine turtles does not affect their ^{15}N values, and the small effect acidification has on ^{13}C values can be mathematically corrected. So there is no need for for prior acidification of a cortical bone (11).

At the moment, there exists no standard protocol for the sequential sampling of cortical bone samples taken from fresh, modern samples for stable isotope analysis (10).

Two methods were tested:

1. Micromilling untreated bone cross sections, and
2. Biopsy coring bone cross sections processed for skeletochronology, for sequentially sampling individual growth layers from marine turtle humerus bones.

A standard protocol has been worked out for sequential bone growth layer sampling for stable isotope analysis. This facilitates a direct comparison in the course of future studies.

The micromilling sampling technique on untreated bone cross sections is recommended. This technique facilitates a higher precision sampling of growth layers that were not affected by chemical processing, and minimized sample handling, thereby reducing the chance for contamination (10).

A study has been presented that compared three different collagen extraction methods commonly used in isotope laboratories conducting dietary studies (12).

The differences in ^{13}C and ^{15}N collagen quality and collagen yield were evaluated. Using the method of Deniro and Epstein (13) with NaOH as cleaning agent, will give ^{13}C values that are on average ± 0.032% more positive than using the ultra-filtration method (14,15). The third method, which is a modified version of the second method, excludes the ultra-filtration step. This method results in ^{13}C values that are in between the other methods. No significant difference in ^{13}C values were found (12).

The method of stable isotopes has also been criticized: It has been found that stable isotopic ratios depend on sufficiently well-preserved molecules and also several collagen quality criteria have to be fulfilled (16). For a better understanding of the chemical degradation of bone collagen, experimentally degraded modern bones and several archaeological human bones were investigated. In the course of the experimental degradation, alterations of isotopic ratios were paralleled by altered collagen quality criteria. The contrary was found in the case of the archaeological specimens. This implies that the commonly used collagen quality criteria may be insufficient and do not guarantee that stable isotopic values of the gelatine extracts will still represent the original biological signal (16).

4.3.3 *Amino Acid Analysis*

The results of amino acid analysis of gelatin collected using two different methods of extracting the organic component in bone have been also discussed. The gelatin which resulted from the demineralization of bone pieces in weak acid, i.e., 1% HCl, consistently showed a pattern of an amino acid composition that is similar to collagen (17).

In contrast, the gelatin which resulted by hydrolyzing demineralized bone powder was unreliable in the retention of protein. Thus, using the former method, rather than carbon-to-nitrogen ratios or the percentage of yield, appears to be the best procedure of material analysis to get a biological signature.

Bone samples appropriate for trace element and stable isotope analysis could be identified even though it was not always possible to identify the diagenetic process (17).

4.3.3.1 *Fossil Dating by Amino Acid Racemization*

A method for determining the age of fossils has been presented. The method evaluates the racemization of isoleucine and other protein amino acids (18,19).

By the fast racemization of amino acids, followed by ion exchange column chromatography separation with a chiral silica gel layer, the age of bones between 5,000 and 50,000 years old could be determined with an error of 5% for *D*-allo-isoleucine and some 15% error for other amino acids. The data indicate that bone samples have an approximate age of 1,000 *y*, by the application of sulfur-containing amino acids (18).

4.3.4 *Ancient DNA*

The study of deoxyribonucleic acid (DNA) extracted from archaeological specimens is an exciting avenue of research that can provide unique evidence for addressing archaeological questions (20). Some notable case studies have been described. Ancient DNA differs from modern DNA. Thus, usually only a few specific genomic markers are targeted. Protocols have been developed explicitly for the study of ancient DNA. Furthermore, ancient DNA recovery and the analysis of genetic material from paleontological, archaeological, and museum specimens have been described in detail (21). Also, DNA from plant compression fossils has been reviewed (22).

The study of ancient DNA offers the possibility of following genetic change over time. However, the field is plagued by a problem which is unique in molecular biology—the difficulty of verifying results by reproduction (23).

Some of the reasons for this are technical and derive from the low copy number and damaged state of ancient DNA molecules. Another reason is the unique nature of many of the objects from which DNA is extracted. Some methodological approaches with which these problems can be alleviated have been described. These approaches should ensure that the results are scientific in the sense that they can be reproduced by others (23).

DNA specialists recommend that the recovery of bones for DNA analysis follow a strict protocol. This includes handling with gloves and not washing or freezing the samples (24,25).

It has been shown that some negative effects can be overcome in many cases (26).

Also, the issues of DNA analysis in paleopathology have been described (27). The discipline of paleopathology, how it has developed, how it is studied, and also its limitations have been described in detail (28).

Studying diseases has a long history and has rapidly developed over the last 40–50 years with the development of methods, in particular with the analysis of ancient pathogenic DNA. For example, this analysis allows the exploration of how strains of the bacteria causing tuberculosis have changed over time (28).

However, there remain issues to be addressed about the quality of the resulting publications (29). Published papers that have described the detection of pathogenic DNA in human skeletal and mummified remains from archaeological sites have been evaluated. The main problematic issues in relationship to standards developed in molecular biology have been discussed. Some 65 papers published between 1993 and 2006 were surveyed and their quality was assessed. Of particular note was the high number of papers that did not acknowledge the use of even a basic contamination control (90%) or procedures to validate results independently (85%). Thus, it has been demonstrated that attention to contamination control and authentication of results is needed to increase confidence in DNA analysis in palaeopathology (29).

4.3.5 *Dentin Layers*

Since dentin is deposited incrementally as growth layers and is neither resorbed nor remodeled once formed, it can reflect the isotopic ratios at the time of its synthesis. Variations in isotopic ratios in sequential dentin layers allow for the reconstruction of ontogenetic dietary changes at the individual level from which population level data can be drawn for comparison of diets across different time periods and geographic areas.

Several techniques have been used to sample annual dentin layers in mammalian teeth for stable isotope analysis. However, tooth size in smaller animals and the conical arrangement of the dentin layers have constrained precision of subsampling and collection of adequate sample from each annual growth layer.

An alternative subsampling technique using teeth from Risso's dolphins (n = 15) has been tested that involved cutting out the annual dentin growth layer groups from 300 to 500 μm longitudinal sections from one half of a demineralized tooth, and comparing the results to those obtained using a standard micromilling process on the other half of the same tooth.

The subsamples were analyzed for elemental carbon and nitrogen content and for stable carbon and nitrogen isotopes. Subsamples obtained from cutting out the growth layer groups showed more consistency with percent amounts of carbon and nitrogen and atomic C-to-N ratios that were significantly different from those obtained by micromilling.

Consequently, the ^{15}N and ^{13}C values differed significantly between methods with values from the former method being more concordant with the expected variations in the early years of Risso's dolphins. Deviations in ^{13}C values in the micromilled subsamples were large enough to create possible errors in interpretation of the dietary sources (30).

Cutting out the dentin layers reduced sample processing time by 90% and yielded ca. 10% more collagen than micromilling. These results suggest that cutting out the annual dentin layers can produce greater yield of samples of better collagen quality with a much shorter processing time than the micromilling process and is, therefore, an effective method to subsample small mammalian teeth (30).

4.3.6 *Evolution of Diseases*

A comparative study of the immunoglobulin G content of the bones of five prehistoric children showed the lowest immunoglobulin G content in a child who suffered from chronic scurvy. Ancient bone proteins were also separated using two-dimensional electrophoresis. This technique makes fractionation of the complex protein mixtures of extracellular bone matrix more reproducible.

Bone retains a chemical memory of earlier metabolic stimuli in its configuration of collagenous and noncollagenous proteins. In combination with the results of a microscopic examination of ancient bone, it should be possible to obtain more reliable information on the history and the evolution of diseases, based on the analysis of intact proteins (7).

4.3.6.1 *Bone Diagenesis*

When a bone is exposed to the burial environment it may experience structural changes induced by multiple agents (31). Studying the post-mortem alterations is an important issue in order to differentiate decomposition phenomena from normal physiological processes or pathological lesions, as well as to assess the quality of the bone tissue.

Microscopy is an important tool to evaluate the integrity of the bone microstructure and it provides significant data on long-term bone decomposition. In a special study, eight archaeological human bone sections and ten human bone sections retrieved from an identified skeletal collection were selected for analysis under plane light and polarized light. The goal of this study was to analyze the impact of diagenesis and taphonomy on the bone microstructure, as well as on the differential diagnosis of pathological conditions (31).

The results showed that the microscopy approach to bone tissues contributed materially as an aid in the detailed description of the main diagenetic changes observed. However, the study showed that gross inspection does not provide a realistic assessment of bone tissue preservation, which can impact the characterization of lesions present and subsequent disease diagnosis.

Therefore, histological techniques are recommended if the goal is to comprehend tissue integrity and its association with decomposition or disease (31).

4.3.7 *Paleodietary Studies*

Biogeochemical approaches to paleodietary analysis have been reviewed (32). This includes bone chemistry, food and history, as well as the cultural effects of stable isotope variations of past human European populations and the details of the analytical techniques used therein.

Also, the bone chemistry and trace element analysis have been described in a monograph (33).

Prehistoric diets may be reconstructed from the chemical and isotopic composition of human bone (34).

Bone chemistry paleodietary studies are emerging as important research areas in archaeology, biological anthropology, and paleontology (35). With appropriate controls, the inorganic and organic chemical composition of bones and teeth can provide information about past diet and habitat use. Chemical signatures are used to address individual dietary variability in early hominid fossils, paleontological remains, and more recent human and faunal populations.

Several methods for assessing the preservation of the organic and inorganic components of bone have been discussed (17). Some aspects of bone composition, e.g., specific trace element concentrations and stable isotope ratios, can provide information on the diet and health status of prehistoric and human fossils. Such information is valuable in order to understand the evolution of human adaptation but, prior to the analysis of bone, it is necessary to ensure that the biological signature has not been erased during burial or following excavation.

Comparatively quick methods of scanning sets of bones have been discussed in order to eliminate methods that are not appropriate for analysis (17). The results of powder X-ray diffractometry have been presented. Also, a survey of thin sections of human bones recovered from a coastal site in Georgia, USA, to assess the preservation of biogenic bone mineral and histological structure has been discussed.

The bone samples that displayed a lack of histological structure also showed X-ray diffraction patterns which deviated markedly from that of fresh bone. It was concluded that a survey of thin sections of a small number of samples in any skeletal population is a rapid, dependable means of eliminating bone which has been altered diagenetically (17).

Other studies of bone chemistry have been used to reconstruct the habitat of ancient populations (36). The problems associated with changes in bone crystallinity and the process of fossilization have been assessed. The results suggested that the older the bone the higher is its crystallinity and the sharper is its X-ray diffraction pattern. This phenomenon could be used as a method for estimating the degree of fossilization and could also be developed as a method for relative dating.

The technique is relatively rapid and economical to apply, uses equipment that is widely available and is practically nondestructive, so it could be used on precious bones such as those of human fossils (36).

4.3.7.1 Strontium Analysis

Isotopic methods are widely used in archaeology to investigate paleodietary issues (37).

A method to identify trophic level in archaeological human populations and to investigate paleodiet has been developed. Strontium isotope compositions $^{88}Sr/^{86}Sr$ have been shown to vary in a mass-dependent manner with increasing trophic level. This fact can elucidate paleodiet in archaeological human populations.

Mass-dependent strontium isotope data from tooth enamel and bone have been presented from individuals buried during the Late Intermediate Period (1000–1300 AD) in the large cemeteries of Chiribaya Alta, Chiribaya Baja, San Gerónimo, and El Yaral in the Ilo and Moquegua Valleys of southern Peru.

These radiogenic strontium isotope data $^{87}Sr/^{86}Sr$ and light stable isotope data (^{15}N and ^{13}C) from the same individuals were used to investigate the geologic variability in strontium sources as well as marine food consumption among the Chiribaya. The results showed the utility of measurements of strontium isotope fractionation as a fine tool for archaeological investigation of paleodiet.

Importantly, this technique can be used to generate paleodietary $^{88}Sr/^{86}Sr$ and paleomobility $^{87}Sr/^{86}Sr$ data from the same specimen, thus minimizing the destructive analysis of invaluable archaeological material, and provides a way to examine paleodiet through hydroxyapatite, which is particularly important when collagen is only poorly preserved (37).

A comparative analysis of trace element data of human skeletal material from four archaeological sites has been done (38). The state of histomorphological preservation of bone, which is a measurement for the degree of its diagenetic alteration, closely corresponds to the Sr/Ca and Ba/Ca ratios at an intrapopulation level.

So, this study challenges the validity of dietary studies that are based on the commonly used procedures for sample preparation and subsequent evaluation of strontium and barium concentrations (38).

4.3.7.2 *Paleodiet Reconstruction*

The isotopic compositions of the tissues of animals and humans could be determined by the proportions of the various nutrients which they consume (39). This allows determining how much of each of their available foods were actually consumed over various parts of their lifetime.

This method was applied to ancient humans by the analysis of their bones, teeth and, rarely, desiccated softer tissues, such as hairs and skin. The variations in ^{13}C and ^{15}N of known classes of nutrients occurring in plants and animals have been analyzed.

The chemical species whose isotopic composition can be analyzed are collagen, bone mineral, and lipids (39). Some deviation in the ^{15}N content may also arise from variations in ^{15}N in soil particulate matter in the sea. This effects an aridity on the nitrogen balance in terrestrial animals.

Examples have been presented of the variation in ^{13}C due to the consumption of maize and in ^{15}N due to trophic effects in prehistoric hunter-gatherers (39).

4.3.7.3 *Preservative-Treated Samples*

Collections of human bone from archaeological contexts are usually treated with resins, glues, and varnishes to consolidate and strengthen them. The use of such treated bone in chemical analysis for dietary reconstruction has been evaluated. The results from a study of a museum collection that had been treated with a common consolidant, Alvar, have been presented. Treatments to remove Alvar from bone were successful. In addition, for one collagen preparation technique, such pretreatment was not necessary (34).

4.3.7.4 *Dietary Variability at Ancient Pompeii*

The excellent preservation of human remains and associated foodstuffs at Pompeii, due to the sealing of archaeological deposits by volcanic ash from the catastrophic eruption of Vesuvius in 79 AD, provides an unusual opportunity to employ stable isotope dietary reconstruction using human and faunal bone, bread remains and a range of domestic plant foods derived from closely associated deposits.

Stable carbon and nitrogen isotope results have been presented for a sample of human and faunal bones and food remains excavated from Pompeii (40). The well-preserved organic remains are a valuable resource to examine ancient way of life associated with a Roman city during a distinct period of time associated with the eruption of Vesuvius in 79 AD. The isotopic results indicate human diets that are consistent with the archaeological and documentary evidence for diverse dietary intake for all residents of Pompeii, consisting of a range of cereals, fruits, vegetables, seafood, and meat from domestic and wild animals.

Bone collagen stable carbon and nitrogen isotope results for a range of domestic animals and foodstuffs recovered from the Pompeii archaeological site are presented in Table 4.1 as a reference sample for human palaeodietary inferences.

The correction of stable carbon isotope values by −0.2% is in relation to changes in isotopic composition of atmosphere associated with burning of wood and fossil fuels.

Carbon isotopic values of −2.81 to −2.45% for domestic plants and local grasses indicate a C3 photosynthetic pathway. On the other hand, nitrogen isotope values for plants are more variable, with a range from 0.06 to 0.81%.

Males had similar animal protein intakes to those of females, but they ingested significantly greater amounts of seafood than females.

Table 4.1 Stable carbon and nitrogen isotope results for domestic animal bone collagen and foodstuffs (40).

Sample	n	$^{15}N/[\%]$	$^{13}C/[\%]$	$^{13}C/[\%]$ corrected
Dog	1	0.85	−1.92	−2.12
Pig	1	0.70	−2.02	−2.22
Horse	1	0.42	−2.04	−2.24
Sheep/Goat	1	0.35	−2.10	−2.30
Cattle	1	0.34	−2.03	−2.23
Erba Grass	1	0.18	−2.66	−2.86
Grapes	1	0.81	−2.64	−2.84
Peas	1	0.20	−2.61	−2.81
Barley	1	0.08	−2.37	−2.57
Lentils	1	0.50	−2.30	−2.50
Bread	1	0.71	−2.17	−2.37
Garum	2	0.49	−1.22	−1.42

The great range of dietary variability among and between males and females indicated by the isotopic data suggests the existence of past dietary distinctions based on their social role or social class.

In comparison to other isotopic studies in the region, human diets at Pompeii are similar to those of Danish farmers and coastal Greek colonies in southern Italy (Metaponto), but included greater amounts of marine protein. The carbon isotope values for domestic animals indicate a C3-based terrestrial diet, while the nitrogen isotope values distinguish herbivores from omnivores and carnivores (40).

4.3.7.5 *Dietary Behavior of Extinct Giant Tortoises*

Taphonomic research of bones can provide additional insight into a formation of a site and its development, the burial environment and ongoing post-mortem processes.

Femur bone samples of 30 tortoises (*Cylindraspis*) from the Mare aux Songes site (Mauritius) have been studied histologically, assessing parameters such as (41):

- Presence and type of microbial alteration,

- Inclusions,
- Staining or infiltrations,
- Degree of microcracking, and
- birefringence.

The absence of microbial attack in the 4200-year-old Mare aux Songes bones suggests that the animals rapidly entered the soil whole-bodied and were sealed anoxically.

However, they suffered from biological and chemical degradation, i.e., pyrite formation or oxidation, mineral dissolution and staining, in comparison to changes in the hydrology of the site.

In addition, carbon and nitrogen stable isotopes were analyzed to get information on the feeding behavior of the animals. The results show narrowly distributed ^{13}C ratios. This indicates a terrestrial C3 plant-based diet. A wide range in the ^{15}N ratios was discovered. This is most likely related to the tortoises' drought-adaptive ability to change their metabolic processes, which can affect the ^{15}N ratios.

Zooarchaeology by mass spectrometry (ZooMS) collagen fingerprinting analysis successfully identified two tortoise species (*C. triserrata* and *C. inepta*) in the bone assemblage, which, when combined with stable isotope data, revealed significantly different ^{15}N ratios between the two tortoise species (41).

As climatic changes around this period resulted in an increased aridity in the Mascarene Islands, this could explain the extremely elevated ^{15}N ratio in the data.

The endemic fauna was able to endure the climatic changes 4200 *y* ago, although human arrival in the 17th century changed the original habitat to such an extent that it resulted in the extinction of several species (41).

4.3.8 Electron Spin Resonance Dating

Electron spin resonance (ESR) is a valuable tool for dating eutherian teeth from archaeological, paleoanthropological, and paleontological sites ranging in age from 10,000 to 5,000,000 years (42).

Because the tooth enamel is the hardest tissue in the body, teeth are frequent components of fossil assemblages. The ESR signal stability in enamel exceeds 100 millions of years. In eutherian teeth, the ESR signal does not depend on the species.

Metatherian, reptilian and chondrichthyian teeth have been tested in order to see whether the ESR method can be extended to additional environments and time periods.

In diprotodontid marsupial teeth, the ESR signal appears identical to that in eutherian enamel. However, the crocodylian teeth that have been tested contained significant iron concentrations, which interfered with the ESR signal. Their fossilization in fluvial sediment may have caused the iron contamination.

Although shark teeth contain enameloid rather than enamel, the ESR signal morphology differs only slightly. However, the signal stability is reduced and the calculated ages have been found to be much younger than expected (42).

4.4 Conservation Methods

4.4.1 *Interventive Conservation Treatments of Pleistocene Bones*

The Pleistocene record can be simultaneously studied from different disciplines within the fields of archaeology and palaeontology (43). Each of these disciplines seeks slightly different information and makes use of different methodologies. These differences may change the goal of the conservation treatments, affect the degree of intervention or require the limited application of certain techniques and materials.

Basic reasons have been discussed as to why some researchers are requesting a closer focus on recovering the original appearance of the bones while others require their taphonomic modifications to remain intact. Further research into the needs of archaeopalaeontological studies and into adjusting conservation aims and methods to meet those needs would help to maximize the recovery and preservation of the information that fossils contain (43).

4.4.2 *Large Fossils*

Large, heavy fossils, such as mammoth remains, add difficulty to the work, as these fossils generally require special supports (e.g., poly(urethane) jackets) starting from the initial lifting, but also during subsequent conservation processes as well as during storage or display (44).

4.4.3 *Micropreparation*

Many vertebrate and invertebrate fossils are quite small and need to be prepared with the aid of a microscope (45). Of course, many of the techniques used for preparation of large fossils can be modified to be used with a microscope.

If a fossil is very delicate it may need to be imbedded in a temporary support such as Carbowax or cyclododecane. Carbowax™ is a water soluble wax. It is a type of poly(ethylene glycol) with molecular weights of 1,450–8,000 *Dalton*. Cyclododecane is a paraffin that sublimates at room temperature.

The main difference related to Carbowax is that since cyclododecane sublimates at room temperature, it is not necessary to wet the fossil or the matrix to remove it. This is very useful for fossils that are water sensitive or in water-sensitive matrix. It will solidify much faster, allowing only a small window of time when it is of useful viscosity. Cyclododecane has a slightly unpleasant odor, whereas Carbowax is odorless.

The basic concept for both conservation materials is very similar. The preparator makes a small mount that will hold the fossil in a bed of wax, allowing the stabilization of the fossil while it is being worked on it. The wax gives support to thin and delicate bone, when it is being worked on.

After the matrix is sufficiently removed, the fossil and mount are allowed to sit in either water or air to remove the waxy support. Fossils that are water sensitive should not be treated with Carbowax, as removing it involves getting the fossil wet. Carbowax and cyclododecane can also be painted onto one side of a delicate fossil to give it stability as the other side is prepared (45).

In some situations, the matrix may be water sensitive. If this is the case, the sample should not be soaked until all matrix is removed. It is often possible to prepare the specimen out of the mount in its own little cocoon of Carbowax, then reset in into the mount with another dose of Carbowax.

Many small fossils arrive from the field with a thin coating of some sort of preservative, usually a polymer, such as poly(vinyl acetate). At the microscopic level, these coatings can often be removed simply by applying gentle pressure to them, separating them from the bone or tooth below, or by using sharp chisel end of the probe to lift the plastic off the specimen.

A cyanoacrylate polymer can be much more difficult to remove and should be avoided as a field preservative. Any small cracks that appear under the microscope should be stabilized as soon as possible. Thin cyanoacrylate glues were found to be best for this, as they easily wick into a crack. The synthesis of a cyanoacrylate compound is done by the Knoevenagel reaction. This is shown in Figure 4.1.

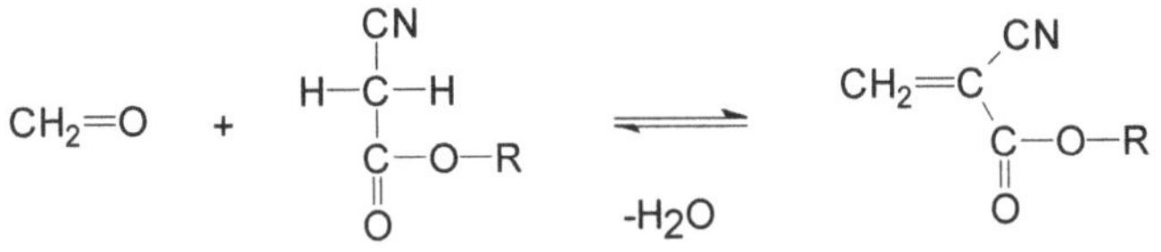

Figure 4.1 Synthesis of a cyanoacrylate.

Other glues do not wick as well. A technique for applying minute amounts has been addressed as the scratch technique.

Here, a small drop of glue is placed onto a scrap of paperboard. With a sharp probe, the drop is scratched, picking up a little wad of paperboard fiber and glue. This wad can then be applied to the crack by lightly touching the wad to the crack. Then the glue will be drawn into the crack. For extremely small quantities of glue, one may be able to scratch the glue drop and find a microdroplet of glue at the end of a paperboard fiber. This is much more difficult to do than a simple fiber scratch. With a simple fiber scratch, the amount of glue may still be too much (45).

Gelatin capsules can make excellent temporary holders for small fossils, especially for broken pieces that will be glued later on. Very small labels can be made inside the gel cap with the specimen. Empty gel caps can be bought at health food stores and pharmacies.

Caffeine and sugar gelatin capsules should be avoided when working under the microscope. Both these products can contribute to manual unsteadiness which will be multiplied under the microscope, making detailed work unnecessarily challenging (45).

4.4.4 *Reaction Adhesives for Fossil Preparation*

Fossil preparators have a wide range of adhesives to choose from. It is often difficult to select the most suitable adhesive for a specific task. The adhesives that preparators use can be divided into two basic categories (46):

1. Solution adhesives, and
2. Reaction adhesives.

Both types of adhesives share some basic characteristics. However, solution and reaction adhesives differ fundamentally in the manner in which they set or solidify (46).

In contrast to solution adhesives, reaction adhesives are often fairly complex formulations, with ingredients that can vary from one manufacturer to another. Commonly used adhesives are summarized in Table 4.2. Poly(vinyl butyral) is shown in Figure 4.2.

Table 4.2 Solution adhesives and reaction adhesives (46).

Solution adhesives	
Brand name	Chemical name
Paraloid B-72	Ethyl methacrylate copolymer
Acryloid	Ethyl methacrylate copolymer
Butvar B-76	Poly(vinyl butyral)
Butvar B-98	Poly(vinyl butyral)
McGean B-15	Poly(vinyl acetate)
Vinac	Poly(vinyl acetate)
Reaction adhesives	
Brand name	Chemical name
Devcon	Epoxy resin
Epo-Tek	Epoxy resin
Aron Alpha	Cyanoacrylate resin
Paleobond	Cyanoacrylate resin

Selecting the most appropriate adhesive for the desired task is important for successful fossil preparation. In practice, no two fossils are exactly alike. Furthermore, an experienced preparator

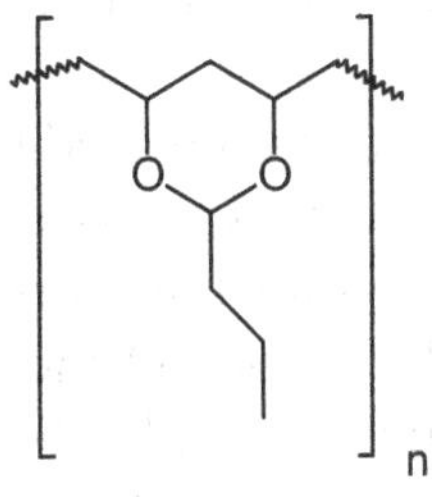

Figure 4.2 Poly(vinyl butyral).

is often faced with new challenges that require them to reevaluate an old approach or to develop new solutions.

There is no single adhesive that works best in every preparation situation. The adhesives listed in Table 4.2 can be subdivided into two basic groups according to whether they set or dry. Solution adhesives are set by evaporation of a solvent. Reaction adhesives are set by a chemical reaction (46).

The organic solvents commonly used for fossil preparation can easily redissolve solution adhesives since they are held together with weak secondary bonds. On the other hand, reaction adhesives are held together by strong primary bonds. However, there are some situations where reaction adhesives can be softened with solvents and removed successfully when they are used in very small amounts. But in almost all cases, removing or reducing reaction adhesives will require more time, effort, and far greater risk to the fossil (46).

Resolubility can be advantageous as field work often requires a temporary application of consolidants or coatings. Resolubility is important in both long-term and short-term or temporary applications.

There are only some rare emerged cases where resolubility may be undesirable, such as when smaller, more delicate joins could dissolve by accident if the surface was exposed to a solvent during cleaning or application of a coating in preparation for molding (46).

Barrier layers of solution adhesives can be applied prior to the application of an epoxy in order to allow for greater reversibility of the join in the future. It has been shown that this technique, if executed properly, can be used without negatively impacting the strength of a join (47).

Soluble barrier layers can also be used to increase the reversibility when reaction adhesives, such as epoxy putties, are used to fill gaps.

When adhesives are used for long-term applications their aging properties should be well understood and proven to be good. A poor aging property of an adhesive can lead to a variety of undesirable results, including shrinkage, distortion, embrittlement, decreased solubility, and darkening or yellowing over time. Damage from poor aging of adhesives can be found in many fossil collections, often including join failures, and embrittled, lifting coatings that have damaged the surface of a bone (46).

The aging behavior of cyanoacrylates has not been deeply investigated. This is partially attributable to the fact that they have not been used widely in the field of art conservation, which has initiated many of the previous assessments of adhesive stability applicable to fossil preparation. A study of cyanoacrylates has shown that there are still many unanswered questions regarding their stability and that contact with some fossils may accelerate degradation of cyanoacrylates (48).

The unique properties of cyanoacrylates, such as their superior ability to penetrate into hairline cracks and their fast set time, may override questions about their aging in some instances (46).

In the conservation of fragile fossil bone material, the impregnation by a solvent-borne consolidant is often required. Understanding the mode of penetration of consolidants into fossil bone is of crucial importance. It is governed by a variety of factors. Neutron imaging is a powerful tool to monitor and visualize the penetration nondestructively. The consolidation of a vertebrate fossil from the Maastrichtian of southeast Netherlands was imaged at the High Flux Reactor facility at Petten, The Netherlands. The analysis showed that the current conservation practice results in a sufficiently deep and isotropic penetration (49).

4.4.5 *Histological Core Drilling*

Paleohistology studies the internal microscopic structure of fossilized vertebrate tissues like bone, ossified tendon, eggshell, teeth and dermal scutes (50).

The previous practice of bone histological studies involves the destructive sectioning of complete bones, teeth, scutes or other preserved hard tissues.

Therefore, museum and collection curators have been reluctant to give up rare or type specimens to these destructive methods. This has limited the paleohistological studies in the last century mainly to isolated and or fragmented specimens, with a few exceptions.

Thus, a less destructive method has been developed. The most important advantage of the histological coring method is that it is less destructive. The most important advantage of the histological coring method described here is its less destructive nature compared to the cross-sectioning of whole bones. Comparable to a biopsy on living tissue, only a small fragment of the tissue is taken, and the morphology of the specimen is preserved.

The histological coring method involves drilling a small core at a standardized location on the bone chosen for study (50). The core is collected using diamond-studded drill bits mounted on a standard household power drill.

The drill is stabilized in a drill press to control the vertical drilling speed. Long bones are preferably sampled because of their abundance and relatively simple growth pattern and morphology, but any vertebrate hard tissue can be sampled. Using an appropriately sized drill bit means any specimen ranging in size from small to extremely large can be sampled.

The cores frequently shear off and in most cases are not recovered in one piece. When a core segment has sheared off in the drilling process, it is important to stop drilling and retrieve it, or it may be damaged. However, core segments that have broken off are usually easily glued back together. Usually, a cyanoacrylate glue is used, such as Paleobond®. Cyanoacrylate has the advantage of fast curing and a refraction index close to 1. Also, it will not interact with the epoxy resin used for embedding the cores. However, other glues that will not obscure the thin sections or interact with the resin may also be used.

After a core is obtained, it is processed into histological thin sections or polished sections with standard histological thin sectioning methods.

In comparison to classical histological techniques, the histological coring method is a much less destructive method for sampling vertebrate hard tissues. This new method will therefore allow the continued conservation of rare and valuable specimens, while simultaneously permitting access to unique biological information (50).

Actually, bone histology has emerged as a major source of information on the life history of dinosaurs (51). Sauropod dinosaurs were the largest animals in the history of life.

4.4.6 *Manual Centrifuge for Resin Casting*

The University of Michigan Museum of Paleontology produces casts of fossils in a wide range of sizes (52). The preparation laboratory often casts small fossils, such as mammalian teeth rows, using silicone molds and tinted epoxy resin for casts. A common problem encountered in the casting process is that small air pockets become trapped in the molds when resin is poured into them, resulting in voids in the extremities of the casts. To mitigate this problem, a centrifuge has been developed to force the resin into the molds, thus displacing the trapped air.

In the construction, quarter-inch-thick acrylic (or polycarbonate may be used) was cut to cover the centrifuge to allow the observation of operation, retain flying resin, and to prevent anything from falling into the centrifuge. The wall is attached to the plywood base using plywood cleats cut to match the curvature of the wall. A similar piece was cut to fit the top of the wall, to which the acrylic lid is attached (52).

4.4.7 *Interferences of Conservation Treatments with Subsequent Studies on Fossil Bones*

Most fossil bones involved in archaeological or paleontological studies may undergo conservation or preparation treatments prior to the analysis of researchers (26). Actually, these treatments are often the only way to recover and save these materials, and consequently to allow research work to be successfully completed.

Various publications have warned about the possible negative effects of conservation treatments on the future study of skeletal objects (53,54). Mostly it is suggested that it is best to treat fossils as little as possible and to leave some untreated samples.

Nevertheless, some conservation methods or materials may negatively interfere with subsequent studies. The research specifically directed towards assessing the possible effects of two common conservation techniques has been reviewed. This is mainly cleaning and consolidation.

Mechanical cleaning of vertebrate fossils can be performed with brushes and small metallic tools such as scalpels or needles. Stronger tools are chisels or ultrasonic scalers, pneumatic air scribes, grinding rotary tools, and sandblasting. Mechanical cleaning is useful when liquids cannot be safely introduced onto the specimens. It has the advantage of being nontoxic and residue-free. One drawback of the mechanical methods is that blows or vibrations can cause fragments to be detached or even give rise to serious fractures on the specimens.

Chemical cleaning of fossil bones is basically used to remove calcareous matrices by the immersion of the samples in acidic aqueous solutions such as acetic and formic acid. These organic acids are most often chosen because they cause only a little damage to the bones. This cleaning method may be the safest as it does not imply striking or introducing vibrations into the fossil matrices. Chemical cleaning also has the advantage of being able to remove the intrusive matrix from even the narrowest gaps, where mechanical tools cannot always reach. Nevertheless, acids may also damage the calcium phosphate in bones. Therefore, the protection of bones from excessive exposure to acid is in generally recommended, along with the use of an initial coat with consolidating products. In addition, after the acid treatment, an in-depth washing process is required. In this way the removal of traces of acid or salts that are produced during the treatment can be achieved. Acetone, or alcohol, can be used for various purposes, including the removal of consolidants or grease from the objects.

Other cleaning techniques are the use of ultrasonic baths, the ultrasonic scaler (25), laser ablation (55–58), and photochemical methods.

The effects of these techniques on bone surface topography as well as on the analysis of some components of the bone tissues, such as isotopes and DNA, have been elucidated.

References

1. A.E. Rixon, *Fossil Animal Remains: Their Preparation and Conservation*, Athlone Press, Distributed by Humanities Press, London, Atlantic Highlands NJ, 1976.
2. E. Parsons, *An Introduction to Marine Mammal Biology and Conservation*, Jones & Bartlett Learning, Burlington, Mass, 2013.
3. C. Dodd, *Reptile Ecology and Conservation: A Handbook of Techniques*, Oxford University Press, Oxford, 2016.
4. M.H. Schweitzer, R. Avci, T. Collier, and M.B. Goodwin, *Comptes Rendus Palevol*, Vol. 7, p. 159, 2008.
5. C.D. Wylie, Preparation in action: Paleontological skill and the role of the fossil preparator, in M.A. Brown, J.F. Kane, and W.G. Parker, eds., *Methods in Preparation. Proceedings of the First Annual Fossil Preparation and Collections Symposium*, pp. 3–12, Petrified Forest, 2009. University of Nebraska State Museum.
6. S. Moore-Fay, A new suit for the Dublin pliosaur, in *Symposium of Paleontological Preparation and Conservation*, Dublin, 2008. National Museum of Ireland.
7. T.H. Schmidt-Schultz and M. Schultz, *American Journal of Physical Anthropology*, Vol. 123, p. 30, 2004.
8. P. Semal and R. Orban, *Journal of Archaeological Science*, Vol. 22, p. 463, 1995.
9. M.J. Schoeninger and K. Moore, *Journal of World Prehistory*, Vol. 6, p. 247, 1992.
10. C.N. Turner Tomaszewicz, J.A. Seminoff, L. Avens, and C.M. Kurle, *Methods in Ecology and Evolution*, Vol. 7, p. 556, 2016.
11. C.N. Turner Tomaszewicz, J.A. Seminoff, M.D. Ramirez, and C.M. Kurle, *Rapid Communications in Mass Spectrometry*, Vol. 29, p. 1879, 2015.
12. M.L.S. Jørkov, J. Heinemeier, and N. Lynnerup, *Journal of Archaeological Science*, Vol. 34, p. 1824, November 2007.
13. M.J. Deniro and S. Epstein, *Geochimica et Cosmochimica Acta*, Vol. 45, p. 341, March 1981.
14. T.A. Brown, D.E. Nelson, J.S. Vogel, and J.R. Southon, *Radiocarbon*, Vol. 30, p. 171, 1988.
15. M.P. Richards and R.E. Hedges, *Journal of Archaeological Science*, Vol. 26, p. 717, June 1999.
16. M. Harbeck and G. Grupe, *Archaeological and Anthropological Sciences*, Vol. 1, p. 43, 2009.
17. M.J. Schoeninger, K.M. Moore, M.L. Murray, and J.D. Kingston, *Applied Geochemistry*, Vol. 4, p. 281, May 1989.
18. J. Csapo, I. Pap, and L. Kolto, *Anthropologia Hungarica*, Vol. 20, p. 67, 1988.

19. J. Csapó, Z. Csapó-Kiss, and J. Csapó, Jr., *TrAC Trends in Analytical Chemistry*, Vol. 17, p. 140, 1998.
20. B.M. Kemp, C. Monroe, and D.G. Smith, *Extraction and analysis of DNA from archaeological specimens* in M. Glascock, R.J. Speakman, and R.S. Popelka-Filcoff, eds., *Archaeological Chemistry: Analytical Techniques and Archaeological Interpretation*, Vol. 968 of *ACS Symposium Series*, chapter 4, pp. 78–98. American Chemical Society, Distributed by Oxford University Press, Oxford, New York, 2007.
21. B. Herrmann and S. Hummel, eds., *Ancient DNA Recovery and Analysis of Genetic Material from Paleontological, Archaeological, Museum, Medical, and Forensic Specimens*, Springer, New York, NY, 2012.
22. E.M. Golenberg, *DNA from plant compression fossils* in B. Herrmann and S. Hummel, eds., *Ancient DNA Recovery and Analysis of Genetic Material from Paleontological, Archaeological, Museum, Medical, and Forensic Specimens*, chapter 17, pp. 237–256. Springer, New York, NY, 2012.
23. O. Handt, M. Höss, M. Krings, and S. Pääbo, *Experientia*, Vol. 50, p. 524, 1994.
24. M. Pruvost, R. Schwarz, V.B. Correia, S. Champlot, S. Braguier, N. Morel, Y. Fernandez-Jalvo, T. Grange, and E.-M. Geigl, *Proceedings of the National Academy of Sciences*, Vol. 104, p. 739, 2007.
25. J. Fortea, M. de la Rasilla, A. García-Tabernero, E. Gigli, A. Rosas, and C. Lalueza-Fox, *Journal of Human Evolution*, Vol. 55, p. 353, 2008.
26. L. López-Polín, *Quaternary International*, Vol. 275, p. 120, October 2012.
27. M. Spigelman, D.H. Shin, and G.K.B. Gal, *The promise, the problems and the future of DNA analysis in paleopathology studies* in A.L. Grauer, ed., *A Companion to Paleopathology*, chapter 8, pp. 133–151. Wiley Blackwell, Chichester, West Sussex, 2016.
28. C.A. Roberts, *Anthropological Review*, Vol. 79, p. 1, January 2016.
29. C. Roberts and S. Ingham, *International Journal of Osteoarchaeology*, Vol. 18, p. 600, 2008.
30. F.C. Evacitas, G.A.J. Worthy, and L.-S. Chou, *Journal of Archaeological Science*, Vol. 66, p. 112, February 2016.
31. S. Assis, A. Keenleyside, A.L. Santos, and F.A. Cardoso, *Microscopy and Microanalysis*, Vol. 21, p. 805, August 2015.
32. S. Ambrose and M.A. Katzenberg, eds., *Biogeochemical Approaches to Paleodietary Analysis*, Vol. 5 of *Advances in Archaeological and Museum Science*, Kluwer Academic/Plenum Publishers, New York, 2000.
33. J. Burton, *Bone chemistry and trace element analysis* in M.A. Katzenberg and S.R. Saunders, eds., *Biological Anthropology of the Human Skeleton*, chapter 14, pp. 443–460. John Wiley & Sons, Inc., 2nd edition, 2007.
34. K.M. Moore, M.L. Murray, and M.J. Schoeninger, *Journal of Archaeological Science*, Vol. 16, p. 437, July 1989.
35. F.D. Pate, *Journal of Archaeological Method and Theory*, Vol. 1, p. 161, 1994.

36. A. Bartsiokas and A.P. Middleton, *Journal of Archaeological Science*, Vol. 19, p. 63, January 1992.
37. K.J. Knudson, H.M. Williams, J.E. Buikstra, P.D. Tomczak, G.W. Gordon, and A.D. Anbar, *Journal of Archaeological Science*, Vol. 37, p. 2352, September 2010.
38. A. Fabig and B. Herrmann, *Naturwissenschaften*, Vol. 89, p. 115, 2002.
39. H.P. Schwarcz and M.J. Schoeninger, *Stable isotopes of carbon and nitrogen as tracers for paleo-diet reconstruction* in M. Baskaran, ed., *Handbook of Environmental Isotope Geochemistry*, Vol. 1, pp. 725–742. Springer Berlin Heidelberg, Berlin, Heidelberg, 2012.
40. F.D. Pate, R.J. Henneberg, and M. Henneberg, *Mediterranean Archaeology and Archaeometry*, Vol. 16, p. 127, 2016.
41. L.G. van der Sluis, H.I. Hollund, M. Buckley, P.G. De Louw, K.F. Rijsdijk, and H. Kars, *Palaeogeography, Palaeoclimatology, Palaeoecology*, Vol. 416, p. 80, December 2014.
42. A.R. Skinner, B.A.B. Blackwell, M.M. Hasan, and J.I.B. Blickstein, *Expanding the range of electron spin resonance dating* in M. Glascock, R.J. Speakman, and R.S. Popelka-Filcoff, eds., *Archaeological Chemistry: Analytical Techniques and Archaeological Interpretation*, Vol. 968 of *ACS Symposium Series*, chapter 1, pp. 1–14. American Chemical Society, Distributed by Oxford University Press, Oxford, New York, 2007.
43. L. López-Polín, *Quaternary International*, Vol. 388, p. 199, November 2015.
44. G.G. Beiner and R. Rabinovich, *Journal of Paleontological Techniques Special Volume*, Vol. 13, p. 19, 2014.
45. J. Cavigelli, Micropreparation ... one sand grain at a time, in M.A. Brown, J.F. Kane, and W.G. Parker, eds., *Methods in Preparation. Proceedings of the First Annual Fossil Preparation and Collections Symposium*, pp. 41–52, Petrified Forest, 2009. University of Nebraska State Museum.
46. A. Davidson and S. Alderson, An introduction to solution and reaction adhesives for fossil preparation, in M.A. Brown, J.F. Kane, and W.G. Parker, eds., *Methods in Preparation. Proceedings of the First Annual Fossil Preparation and Collections Symposium*, pp. 53–62, Petrified Forest, 2009. University of Nebraska State Museum.
47. J. Podany, K.M. Garland, W.R. Freeman, and J. Rogers, *Journal of the American Institute for Conservation*, Vol. 40, p. 15, Spring 2001.
48. J. Down and E. Kaminska, Degradation of cyanoacrylate adhesives in the presence and absence of fossil material, in *Journal of Vertebrate Paleontology*, Vol. 25, pp. 52A–52A, Northbrook, IL, 2005. Soc. Vertebrate Paleontology.
49. A.S. Schulp, R. Schouten, L. Metten, A. van de Sande, and A. Bontenbal, *Netherlands Journal of Geosciences*, Vol. 92, p. 177, 9 2013.

50. K. Stein and M. Sander, Histological core drilling: A less destrcutive method for studying bone histology, in M.A. Brown, J.F. Kane, and W.G. Parker, eds., *Methods in Preparation. Proceedings of the First Annual Fossil Preparation and Collections Symposium*, pp. 69–80, Petrified Forest, 2009. University of Nebraska State Museum.
51. P.M. Sander, N. Klein, K. Stein, and O. Wings, *Sauropod bone histology and its implications for sauropod biology* in N. Klein, K. Remes, C.T. Gee, and P.M. Sander, eds., *Biology of the Sauropod Dinosaurs: Understanding the Life of Giants*, Series: Life of the Past, chapter 17, pp. 276–302. Indiana University Press Bloomington, 2011.
52. D. Erickson, Inexpensive and simple construction of a manual centrifuge for resin casting, in M.A. Brown, J.F. Kane, and W.G. Parker, eds., *Methods in Preparation. Proceedings of the First Annual Fossil Preparation and Collections Symposium*, pp. 93–96, Petrified Forest, 2009. University of Nebraska State Museum.
53. J. Johnson, *Bar International Series*, Vol. 934, p. 99, 2001.
54. G. Panagiaris, *Bar International Series*, Vol. 934, p. 95, 2001.
55. M. Cooper, *Laser Cleaning in Conservation: An Introduction*, Butterworth-Heinemann, Woburn, MA, 1998.
56. F. Landucci, R. Pini, S. Siano, R. Salimbeni, and E. Pecchioni, *Journal of Cultural Heritage*, Vol. 1, p. S263, August 2000.
57. F. Landucci, E. Pecchioni, D. Torre, P. Mazza, R. Pini, S. Siano, and R. Salimbeni, *Journal of Cultural Heritage*, Vol. 4, p. 106, January 2003.
58. L. López-Polin, A. Ollé, J. Chamón, and J. Barrio, Lasers for removing remains of carbonated matrices from pleistocene fossils, in M. Castillejo, P. Moreno, M. Oujja, R. Radvan, and J. Ruiz, eds., *Lasers in the Conservation of Artworks*, pp. 477–481, Boca Raton, 2008. Proceedings of the International Conference Laocna VII, Madrid, 2007, CRC Press.

5

Stones

Certain methods and issues for surveying historic buildings have been detailed in monographs (1, 2). The term survey cannotes different meanings. The principal types of survey are (2):

- Measured survey of a building with respect to construction, layout and appearance,
- Building surveys of the structure and fabric of properties in order to assess their condition for the rules of future restoration, and
- Archaeological surveys of material remains in the ground.

Also, the principles and practice of stone conservation have been described in monographs (3–5).

In seawater, stone is often associated with shipwrecks. Stones have been carried as ballast materials in ships. Also, the deterioration of stones, among others, such as ceramics and glass, has been reviewed (6). Fortunately, ceramics, glass and stone do not deteriorate rapidly in the course of contact with seawater. Mostly a physical deterioration occurs due to mechanical damage.

Stones can be classified into three major categories (6):

1. Igeneous or magnetic rocks, which crystallize from a hot silicate melt,
2. Sedimentary or layered rocks, which are formed by the concentration of debris, and
3. Metamorphic rocks that are derived from igeneous or sedimentary rocks, which have been recrystallized due to pressure or temperature.

The average solubility of some stones in water is shown in Table 5.1.

Table 5.1 Solubility of stone types (6).

Stone type	Average solubility/[*ppm*] Fresh water	Sea water
Calcite	40–85	66
Aragonite	40–85	66
Dolomite		50
Siderite	10–25	
Gypsum	2000	6000
Alabaster		5000–6000

It can seen be that sulfate rocks, such as gypsum, are more soluble than carbonates.

5.1 Deterioration Processes

5.1.1 Biological Deterioration

Great attention has been recently paid to the surface properties of building materials. These allow control of solar gains of the building envelope and overheating of buildings, including urban areas. Deterioration phenomena of building materials due to biological aggression can quickly damage the solar-reflecting roof surfaces and thus sharply increase solar gains, discomfort, air-conditioning costs and waterproofing degradation. The same deterioration problem has a deleterious effect on cultural heritage, ruining its huge historic and artistic value (7).

An overview of the different organisms that affect the surface of most used building materials needs to be done to support the design of new building materials with long-lasting surface properties and to find a way to preserve cultural heritage.

Artificial aging has been investigated to find out what actually happens in nature after months or years. This can be compressed in a very short time by forcing the growth of microorganisms through strict control of different conditioning factors.

Both natural and artificial aging have been described to provide a comprehensive idea of what is necessary to study in a complete way biological aging protocols on building materials (7).

Biological organisms that affect building materials are summarized in Tables 5.2 to 5.4.

Table 5.2 Cyanobacteria and algae affecting building materials (7).

Concrete, Stone, Brick, Mortar, Limestone Algae	
Chlorophyta	*Apatococcus, Bracleacoccus, Chlamydomonas, Chlorella, Chlorococuum, Chlorokybus, Chlorosarcina, Clorosarcinopsis, Cosmarium, Desmosoccus, Ecdysichlamys, Haematococcus, Fiedmannìa, Geminella, Klebsormidium, Leptosiroid, Muriella, Myrmecia, Neochloris, Oedogonium, Palmellosìcoccus, Pleuratrum, Protococcus, Pseudodendoclonium, Scenedesmus, Stichoccus, Stigeoclonium, Tetracystis, Trebouxia, Trentepohlia, Bacilliariophyceae, Chrysophyceae, Eustigmatophyceae, Rhodophyceae, Xanthophyceae*
Cyanobacteria	*Aphanocapsa, Aphanoteche, Borzia, Calothrix, Chamaesiphon, Chlorogloea, Chroococcus, Gloeocapsa, Microcoleus, Myxosarcina, Nostoc, Oscillatoria, Phormidium, Plectonema, Pleurocapsa, Stigonema, Symploca,Synechococcus, Schizothrix, Scytonema, Tolypothrix*
Metal	
Algae	Hydrogenase-positive *chlorophyta*
Cyanobacteria	*Nostoc, Anabaema*
Painted Surface	
Algae	*Chlorella, Chlorococcum, Eustigmatus, Pleurococcus, Stichoccus, Trebouxia, Trentepohlia, Ulothrix*

5.1.1.1 Biodeterioration of Stone

The biodeterioration of stone is coupled with nearly all environmentally induced degradation processes. The bioreceptivity of stone is described by its structure and chemical composition, while the intensity of the microbial contamination is determined by the referring climatic conditions and the anthropogenic euthrophication

Table 5.3 Fungi affecting building materials (7).

Concrete and Stone
Alternaria, Aspergillus, Aureobasidium, Botrytis, Candida, Cladosporium, Coniosporium, Coniothyrium, Cryptococcus, Curvularia, Epicoccum, Exophiala, Fusarium, Geotrichum, Mucor, Mycocalicum, Paecilomyles, Penicillum, Phialophora, Phoma, Sarcinomyces, Sporobolomyces, Taeniolella, Trichoderma, Udeniomyces, Ulocldium, Verticillium, Capnodiales, Hypocreales
Painted Surfaces
Alternaria, Aspergillus, Aureobasidium, Cephalosporium, Cladosporium, Curvularia, Epicoccum, Exophiala, Fusarium, Helminthosporium, Monascus, Mucor, Nigrospora, Pestalotia, Penicillum, Pestalotiopsis, Stachybotrys, Stemphyllium, Trametes, Trichoderma, Tripospermum, Ulocladium, Verticillum

Table 5.4 Bacteria affecting building materials (7).

Concrete	*Thibacillus, Nitrosomonas, Nitrosovibrio, Nitrobacter*
Metal	Sulfate-reducing Bacteria *Desulfovibrio* and *Clostridium*

of the atmosphere (8). The microflora improves the nutrient- and moisture-restricted growth conditions on building stones by the formation of surface-covering biofilms. Besides the aesthetical impairment caused by the colored biopatina, the biofouling effect even promotes abiotic deterioration processes due to the alteration of the material structure as well as its thermo-hygric properties.

Furthermore, a mechanical pressure due to the shrinking and swelling of the colloidal biofilms might cause a further weakening of the mineral lattice. Acidolytic and oxido-reductive biocorrosion processes complete the biodeteriorating attack of stone acting as a preliminary precursor to the latter formation of detrimental crusts.

Biological infections and the intensity of the biodeterioration processes are strongly influenced by the presence of water. This is caused by material-specific parameters, like porosity and permeability, as well as by environmental conditions (9).

Also, the natural biomass accumulated by photosynthetic microorganisms and anthropogenic pollutants from agricultural and industrial sources can solely serve the nutrient demand of the stone-colonizing microorganisms (10, 11).

So, some possible effects of the microbial contamination on cleaning procedures, protective treatments, as well as biocidal applications have to be considered. A comprehensive overview of the biodeterioration of stone has been presented concerning the relevance for conservation procedures (8).

5.1.2 *Biological Colonization on Ceramics*

Architectural ceramic materials exposed to outdoor conditions are often affected by biodeterioration. This worldwide problem can cause aesthetical, physical and chemical damages on ceramic materials. The distinct biodeterioration processes occurring on the different substrates found on architectural ceramic materials have been reviewed in detail. The major issues are biodiversity, biodeterioration and bioreceptivity (12).

Data with regard to biodiversity found on architectural ceramic materials has been described for three typologies:

1. Bricks,
2. Roofing tiles, and

3. Glazed wall tiles.

Actually, a vast biodiversity has been identified on these ceramic materials, from bacteria to more complex organisms, such as plants.

The highest biodiversity was found on bricks. Also, the colonization of ceramic roofing tiles by lichens has been elucidated. Photoautotrophic microorganisms found on glazed wall tiles represent a considerable biodiversity in this ceramic typology, with many genera in common with those identified on bricks and roofing tiles.

Porosity and surface roughness seem to play a major role in the bioreceptivity. Also, the ceramic microstructure has a strong influence on the resistance to biodeterioration (12).

5.1.3 *Biofilm Formation*

Biofilm formation appears first as the discoloration of the stone surface due to organic pigments, e.g., chlorophylls, carotenoids, and melanins (8). Depending on the type of stone, the uppermost layers are later preconditioned by the enrichment of adhesive epilithic biofilms resulting from both fungal and bacterial growth.

The precipitating salts get encrusted together with airborne particles and chemical compounds, serving as an additional nutrient source for the stone microflora (13).

Then, subsequent physical stresses induced by freeze-thaw changes and salt recrystallizations, as well as biocorrosive and biooxidative processes, continue the weakening and leaching of the mineral material under the superficial crust (14).

Finally, the decomposition of the stone-cementing binders results in the weakening of the mineral structure manifested by sanding or granular disintegration of the stone's uppermost layers, thus opening the space for further microbial contamination (15).

The process is enhanced by repeated wet/dry cycling and leads to an enhanced moisture condensation in the pore system of the stone, thus favoring a microbial growth.

Therefore, secondary diffusion barriers for humidity and gases inside the mineral structure should be established (8).

5.1.4 *Bacterial Carbonatogenesis*

The effects of a biological and bioinspired bioconsolidation treatment have been described (16). The efficacy of both the bioconsolidation treatment and the conventional ethyl silicate treatment have been detailed.

For the bioconsolidation treatments, a sterile M3 P nutritional solution was prepared (1% Bacto Casitone, 1% $Ca(CH_3COO)_2 \times 4H_2O$, 0.2% $K_2CO_3 \times \frac{1}{2}H_2O$) in a 10 *mM* phosphate buffer with pH 8 and applied on a stone surface by spraying.

The bioconsolidation treatment was repeated twice every day during 6 days of the treatment to get a constant dampness. The temperature was kept nearly constant at 17–25°C by the convection of hot air. In order to avoid desiccation and direct impact of sunlight on the stone surface, the treated areas remained covered with insulating foils during the treatment, and also 3 days afterwards, until the stones became fully dry.

It has been observed that the level of consolidation obtained by bacterial carbonatogenesis is similar to that achieved with ethyl silicate. However, 5 months after the treatment, the silica gel which formed on the stone shows extensive cracking, which jeopardizes its long-term effectiveness as a consolidant.

In contrast, neither detrimental effects, nor any reduction in the level of consolidation, have been observed in the case of a bioconsolidation treatment (16).

5.1.5 *Microflora on Building Stones*

Actually, the microflora on building stones is a complex ecosystem which can develop in various ways depending on the environmental conditions and the physicochemical properties of the specific material (17).

Microflora are photolithoautotrophic organisms, such as algae, cyanobacteria, mosses and higher plants, which use sunlight as energy source and release oxygen during photosynthesis.

Chemolithoautotrophic bacteria use inorganic compounds, e.g., ammonia, nitrites, hydrogen sulfide, thiosulfates or elementary sulfur to obtain energy from their oxidation and fix CO_2 from the atmosphere. This results in the release of nitrous acids, e.g., *Nitrosomonas* spp., nitric acid, e.g., *Nitrobacter* spp., or sulfuric acid, e.g., *Thiobacillus* spp. A few bacteria of this group are also capable of growing mixotrophically, as they use organic nutrients for the synthesis of cell components. This behavior is addressed as chemolithomixotrophic.

Fungi use organic substrates as hydrogen, carbon, and energy source. They commonly release complexing biocorrosive organic acids or weaken the mineral lattice by the oxidation of metal cations such as Fe^{2+} or Mn^{2+}.

Lichens represent a symbiotic association between a fungus, also known as the mycobiont, and an alga or cyanobacteria, also known as the phycobiont. The fungus utilizes organic nutrients produced by the algae through photosynthesis. In return, the algae is believed to gain minerals leached from the stone by fungal acids secreted by the hyphae of the lichen. The fungus also protects the alga from harmful environmental conditions such as dessication or toxic chemicals.

Chemolithotrophic microorganisms have often been described in association with damaged inorganic materials (18, 19)

However, recent studies have emphasized the significance of chemoorganotrophic bacteria and fungi, together with photoautotrophs, as the primary microbial colonizers of building stones (20,21)

The respective microorganisms are often present in surface biofilms on stones, which allow them to withstand abrupt changes in environmental conditions, such as temperature, humidity, and osmotic pressure, as well as food shortages. The activity of the primary colonizers preconditions the building for the chemolithoautotrophs and gives rise to the biological succession (22).

5.1.6 *Patina Formation on Mineralic Rocks*

In the course of microbial contamination the physicochemical properties of a mineral substrate can be altered significantly, thus improving their resulting bioreceptivity (8). Microbial biofilms, due to physical and chemical weathering, may develop into surface decay forms. This is called patina and may be the formation of crustal films.

Three different issues of patina types have been classified, such as (8):

1. Film-formation,
2. Surface-corrosion, and
3. Crust-formation.

5.1.7 *Cyanobacteria*

The importance of microbial activity in the alteration and deterioration of stone and concrete walls has often been neglected (23). However, active biofilms are found anywhere there are microorganisms and humidity, and a complex biofilm system can develop on external walls of buildings. Microbiota on stones for building represents a complex ecosystem which can develop in various ways, depending on environmental conditions and the physicochemical properties of the material in question. The colonization of the external surfaces of buildings by microorganisms often causes an aesthetically unacceptable appearance of staining on the stone surfaces.

Organisms that may be present on stone monuments include algae, cyanobacteria, mosses, and higher plants. Also, chemolithoautotrophic bacteria may be present that can release nitrous acid, nitric acid, or sulfuric acid (24,25). In this way, the local pH may go down.

Cyanobacteria can colonize a wide variety of terrestrial habitats. The role of cyanobacteria in the deterioration of surfaces of historic buildings has been the subject of several studies (26,27).

The action of cyanobacteria as deteriogens has been described (23). Also, the traditional and more modern molecular methods used to detect these microorganisms have been discussed.

The development of molecular techniques for the rapid identification of cyanobacteria without the need for culture and isolation is fundamental to extend the knowledge of these communities in biofilms on the surfaces of historic buildings (23).

5.1.8 *Microbial Deterioration of Sandstone*

The microbial deterioration of sandstone from Osirion's Sarcophagus Chamber as affected by rising ground water level, has been investigated to detect the responsible fungi (28).

The following methods were used: Scanning electron microscope (SEM), X-ray diffraction (XRD), and energy dispersive X-ray spectroscopy.

The following fungi could be identified on the deteriorated stone surfaces: *Cladosporium cladosporioides*, *Aspergillus terreus*, *Curvularia lunata*, and *Acremonium falciforme*.

The hyphae penetration of the identified microfungi caused a mechanical exfoliation of the building stone material, as well as changes in color (28).

5.1.8.1 *Types of Corrosive Films*

Film Formation. Film-formation is also addressed as patina, deposit, coating, staining, and chromatic alteration. It may occur on the following types of rocks: Dense or fine grained stones, siliceous sandstones, granite, basalt, slate, limestone and metamorphic rocks, i.e., gneiss, quartzite, or marble. This type shows only a poor penetration of maximal up to 1 *mm*.

The distribution and the type of microflora are superficial or along natural cracks and fissures, mostly unilamellar biofilm, mainly dominated by a phototrophic microflora and also fungi. Typical biodeterioration processes are discolorations by biogenic pigments and biogenic oxidation of mineralic iron or manganese, biofilm formation leading to the enrichment of atmospheric particles and subsequently causing the development of thin-skinned scales, and local biocorrosion, i.e., biopitting due to the microbial excretion of organic acids (8).

Surface Corrosion. Surface corrosion is also addressed as granular disintegration, sanding, pulverization, and erosion. It may occur on the following types of rocks: Coarse-grained, porous stones, tuff, clay-cemented or siliceous sandstones, man-made stones, i.e., brick, mortar, or concrete. This type shows a deep penetration up to 10 *cm*.

The distribution and the type of microflora are a microbial contamination up to 5 *cm* deep and mainly dominated by bacteria. Typical biodeterioration processes are biofilm formation narrowing rock pores, possibly leading to an increase in capillary water uptake, and biocorrosion due to the microbial excretion of inorganic and organic acids (8).

Crust Formation. Crust formation is also addressed as exfoliation, chipping, shales, flakes, and scales. Types of rocks are middle-grained sandstones, clay-cemented and calcareous types of rock, respectively, binding material

This type shows moisture balance in the uppermost layers of the stone of 0.5–2 *cm* deep. Typical biodeterioration processes are discolorations by biogenic pigments and biogenic oxidation of mineralic iron or manganese, biofilm formation that seals the rock pores causing a reduced diffusion of humidity and enrichment of atmospheric particles with a subsequent crust formation, and biocorrosion due to the microbial excretion of inorganic and organic acids (8).

5.2 Analytical Methods

5.2.1 *Analysis of Starch*

The analysis of starch grains can provide important data about the functions of stone grinding tools recovered from early archaeological sites in the humid lowland Neotropics (29). Thus, starch grain analysis has the potential to help in understanding the domestication of tropical plants.

The upper-middle Huai River Valley is located in the transition zone in the middle of eastern China. Previous studies have shown that the past populations in the upper Huai River Valley obtained their plant food through gathering and cultivation of both rice and millet during the Middle Neolithic Period (30). However, for the middle Huai River Valley it is not very clear what methods were used by humans to obtain plant food at that time.

A starch grain analysis was carried out on stone tools and cauldron pottery sherds unearthed at Shunshanji, the earliest Neolithic site in the middle Huai River Valley.

There, ancient starches from *Coix lacryma-jobi, Triticeae, Oryza sativa, Trichosanthes kirilowii,* as well as one unidentified species, have been recovered.

First, the used and unused surfaces of the artefacts were identified by naked-eye observation and distilled water was used to clean each surface separately. After that, liquid samples were gathered from both used and unused surfaces with an ultrasonic brush and kept in separate test tubes. A further extraction was carried out in a laboratory. Pottery cauldron sherds were selected.

Each of the sherds was put into different glass beakers for an ultrasonic bath. Liquid samples from the inner and outer side of the pottery sherds were gathered separately and kept in test tubes for further extraction. Glass beakers were all sterilized at high temperature before use. The liquid samples were treated with 10% HCl and 5% $(NaPO_3)_6$ to get information about the carbonates and other minerals.

Then a heavy liquid was used for the centrifuge procedure. In addition, control samples from laboratory consumables were analyzed. The identification of the ancient starch grains was based on a previous morphological analysis of modern starch grains and on a modern reference collection from most of the common plant species used for food during the Middle Neolithic Period in China.

Both gathering and cultivation of rice took place during the Middle Neolithic Period. *Coix lacryma-jobi* appears to have been the main plant food at Shunshanji. According to the occurrence frequency of the different starch grains it became clear that *Oryza sativa* was not consumed as much as other plant species.

Starch grains were also found on the surfaces of grinding stones, as well as pestles. This indicates that these stone tools were used for food processing (30).

5.2.2 *Residue Analysis*

Residue analysis has been used with improving frequencies within lithic studies to gain more information about the function of ancient stone tools (31). Very often this method has been applied in combination with use-wear analysis to improve the accuracy of the functional interpretation of the samples (32).

In spite of the great progress of residue analysis, several methodological problems have been detected with respect to the identification of the type of residue (33).

Some methodological improvements with respect to residue analysis often present on the surface of stone tools have been presented (31). For example, these residues are skin flakes and modeling clay. The characterization of these materials was done by using both optical light microscopy and SEM.

Microscopic analysis of residues on prehistoric stone tools can be used to infer the use and task associated with artefacts. Tool use residues, mainly animal in origin, have been identified on a wide variety of tool types (34).

Infrared reflectance spectroscopy is an analytical technique for the study of residues on stone tools (35). Using a microscope attachment has the additional benefit of analyzing very small spots directly on an artefact without sample preparation or destruction. The method has been tested with replicated Middle Stone Age stone points used during hunting and butchery experiments on parts of a blue wildebeest (*Connochaetes taurinus*).

5.2.3 *Optically Stimulated Luminescence Dating*

Light-emitting diodes (LEDs) have multiple applications in science. Also, archaeology and earth sciences have benefited with respect to chronology based on the thermoluminescence and optically stimulated luminescence methods of dating artifacts of material culture that are of archaeological and geoarchaeological significance (36).

Blue LEDs are mostly used for stimulation and bleaching of luminescence, in contrast to green and infrared LEDs. Ancient inorganic materials made of geological rocks, i.e., carving, knapping, mixing, firing, have been either fired in antiquity above 450°C or have been sun exposed for a short time (surface luminescence dating). Thus the electron traps are completely bleached for the *zero time* clock to set up.

The luminescence from mineral insulators that are measured is usually derived from either quartz or feldspar sand-sized grains, or unseparated silt-sized grains. There are advantages and disadvantages to using each. Quartz is the most abundant in rocks and lithic artifacts. For quartz one normally uses blue or green excitation and measures the near ultraviolet emission, though these also probe feldspars too.

For feldspar or silt-sized grains one normally uses near infrared excitation and measures the violet emission. Dating using luminescence is made possible by the fact that in natural archaeological and geological environment the decay of natural radionuclides, viz. potassium, and radioactive decay chains, viz. uranium, thorium, along with cosmic radiation, provides a constant irradiation field of α, β and γ ray dose rates.

For this reason, the minerals in the sediment or of an archaeological object are irradiated at a constant rate, and hence acquire a latent luminescence at a constant rate. The latent luminescence is then released upon exposure to heat or light, setting the signal to zero or near zero, whence the trapping process begins again (36).

Events that can zero the preexisting geological or archaeological luminescence are intentional or an accidental exposure to heat of about 400°C or the exposure to daylight, which provides sufficiently energetic photons to induce zeroing.

Authigenic minerals that form *in-situ*, such as sediments, also can be dated (37). These minerals acquire their luminescence because of the establishment of their crystal structure. In the laboratory the same process can be mimicked in the course of the dating procedure (38).

Also, a model has been proposed for estimating the soil development time by subtracting the low molecular weight or pyrolysis volatile ^{14}C age of the uppermost horizon of a soil from the optically stimulated luminescence age of the horizon of the parent material (39).

5.2.4 *NMR Devices in Stone Conservation*

Scientists applying magnetic resonance techniques to cultural heritage are now a quite vast and international community, even if these applications are not yet well known outside this community (40).

Not only laboratory experiments but also measurements in the field are possible with the use of portable nuclear magnetic resonance spectroscopy (NMR) instruments that enable noninvasive and nondestructive studies on items of any size, of high artistic and historical value, as well as diagnosis of their conservation state.

The situation became completely different in the second half of the 1990s, when work on applications of NMR to cultural heritage has first begun, which combined the knowledge of NMR for fluids in porous media at the University of Bologna with the skillfulness of the chemists for cultural heritage of CNR and the University of Florence, and the Safeguarding Cultural Heritage Department of Aosta.

The development of these techniques from the first tentative experiments in the second half of the 1990s until recent results has been reviewed (40).

5.2.4.1 *Portable Devices*

Currently NMR portable devices are being used for a nondestructive in situ analysis of the water content, pore space structure and protective treatment performance in porous media in the field of cultural heritage (41).

It has become a standard procedure to invert T_1 and T_2 relaxation data, i.e., longitudinal (T_1) time distribution and transverse (T_1) relaxation time distribution, of fully water saturated samples to get pore size distributions, but the use of T_2 requires great caution.

It is well known that dephasing effects due to water molecule diffusion in a magnetic field gradient can affect transverse relaxation data, even if the smallest experimentally available half echo time t is used in Carr-Purcell-Meiboom-Gill experiments (41, 42). This procedure is a spin-echo method (43,44).

When a portable single-sided NMR apparatus is used, large field gradients due to the instrument, at the scale of the sample, are thought to be the dominant cause of dephasing. In this paper, T_1 and T_2 (at different t values) distributions were measured in natural (Lecce stone) and artificial (brick samples coming from the Greco-Roman Theater of Taormina) porous media of interest for cultural heritage by a standard laboratory instrument and a portable device.

While T_1 distributions do not show any appreciable effect from inhomogeneous fields, T_2 distributions can show strong effects, and a procedure is presented based on the dependence of $1/T_2$ on t to separate pore-scale gradient effects from sample-scale gradient effects. Unexpectedly, the gradient at the pore scale can be, in some cases, strong enough to make negligible the effects of gradients at the sample-scale of the single-sided device (41).

5.2.4.2 NMR Mouse

Portable unilateral NMR is useful for quality assessment of stone conservation treatments (45). To reduce the water ingress of stone in monuments and statues exposed to weather and pollution and to strengthen degraded stone, coatings from organic polymers or silica are applied to the surface. This modifies the surface wetting properties and reduces the pore size in the upper few millimeters of the stone. For measurement of an NMR signal, the stone needs to be wet (46–49).

A signal can readily be obtained with the NMR mouse, but in contrast to the signal of a fully water saturated stone measured in homogeneous fields, the NMR relaxation time distribution determined in strongly inhomogeneous fields from partially wetted stone cannot readily be interpreted in terms of the pore size distribution.

The wetting properties of the stone and the treatment, as well as the effect of the strong magnetic field gradient on signal attenuation by translational diffusion need to be taken into account. Nevertheless, significant differences between the water signals from the wetted sandstone of window frames, treated and untreated materials, could be observed in a field study (45).

5.2.4.3 Relaxometer

Two NMR devices, a standard imaging apparatus and a portable single-sided relaxometer (with geometry that allows nondestructive *in-situ* relaxation measurements on samples of any dimension), were used to assess the hydrophobic performance of an acrylic polymer (Paraloid B-72) applied as a trichloromethane (chloroform) solution on a biocalcarenite, a Lecce stone (50).

The portable single-sided relaxometer has a geometry that allows nondestructive *in-situ* relaxation measurements on samples of any dimension.

The NMR images and relaxation time distribution functions of samples characterized by different treatments were acquired for different water absorption and measurement sequences.

The samples were analyzed both fully saturated and during the capillary water absorption. The relaxation results have been compared with NMR images of the internal sections of the same samples in the same absorption conditions.

The single-sided device was tuned to detect the signal from a sensitive volume that is only a small portion of the entire sample, located in the first 2 *mm* below the surface. In spite of the instrumental and physical differences of the two devices, the experimental results of relaxometry and imaging are in very good agreement.

The comparison allows the assertion to be made that the single-sided NMR technique is a powerful tool for *in-situ* evaluation of water repellent treatments used for consolidation and/or protection of stone artifacts (50).

5.2.4.4 *Pore Size Distributions*

The NMR relaxation time distributions obtained with laboratory and portable devices, are utilized to characterize the pore size distributions of building materials coming from the Roman remains of the Greco-Roman Theater of Taormina (51). To validate the interpretation of relaxation data in terms of pore size distribution, comparison of results from standard and *in-situ* NMR experiments with results of the mercury intrusion porosimetry has been made.

Although the pore size distributions can be obtained by NMR in terms of either longitudinal (T_1) or transverse (T_1) relaxation time distributions, the shorter duration of the T_2 measurement makes it, in principle, preferable, although the determination of T_2 distributions is not necessarily an easy alternative to finding T_1 distributions.

Among other things, the T_1 distribution is almost independent of the inhomogeneity of the magnetic field, while the T_2 distribution is strongly influenced by it. This study was aimed at answering two questions: what are the validity limits to interpret NMR data in terms of pore size distributions and whether the portable device can successfully be applied as a nondestructive and noninvasive tool for *in-situ* NMR analysis of building materials, particularly those of Cultural Heritage interest (51).

5.3 Conservation Methods

Materials in the built environment are exposed to several agents that promote their alteration processes. This results in features that might be considered detrimental.

There are diverse possible intervention processes that might, however, have unwanted side effects. The main issues related to the struggle against these alteration processes have been reviewed.

The consideration of intervention criteria for strategic considerations on the organization of the intervention is detailed and must include temporal and spatial features of alteration processes, as well as possible interventions on the surroundings of the materials and in the materials, including its replacement.

The long-term effectiveness of these interventions is linked with the global strategy namely in relation to the conditions that promote alteration processes. Some sustainability questions related to the intervention operations are also considered such as the use of toxic substances and the consumption of resources.

Some substances can cause emissions after the application, as in the case of leaching of toxic substances used in biocides (52).

5.3.1 *Changes in Petrophysical Properties of the Stone Surface Due to Past Conservation Treatments*

Petrophysical properties, such as porosity, permeability, density or anisotropy determine the alterability of stone surfaces from archaeological sites (53). Therefore, the preservation of the material is dependent on these properties. Superficial roughness or color may point out changes due to the past alteration processes, both natural processes and conservation treatments.

The application of conservation treatments may vary for some of these properties, forcing the stone surface to readapt to the new conditions, which could generate new processes of deterioration.

Changes resulting from the application of consolidating and hydrophobic treatments on stone materials from the Roman Theater (marble and granite) and Mitreo's House (mural painting and mosaics), both archaeological sites from Merida (Spain), have been analyzed. The use of portable field devices allows performing an analysis both on site and in the laboratory, and the comparison of treated and untreated samples. The treatments consisted of synthetic resins, consolidating materials, such as tetraethoxysilane, cf. Figure 5.1, and hydrophobic products. The results indicate that undesirable changes may occur, with consequences ranging from purely aesthetic variations to physical, chemical and mechanical damages (53).

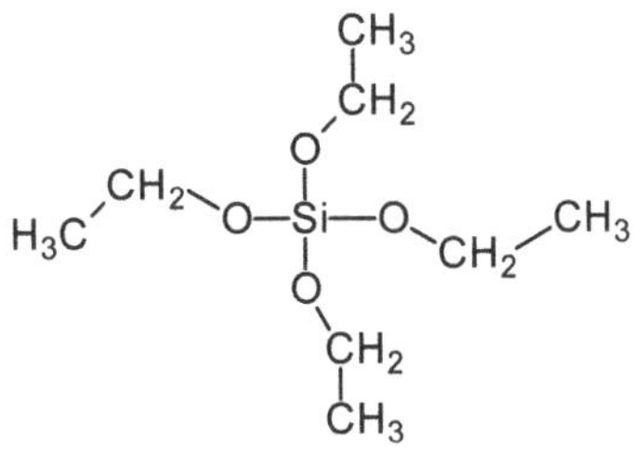

Figure 5.1 Tetraethoxysilane.

5.3.2 *Conservation of Lime*

Both consolidants and protective treatments for the conservation of deteriorated limestone and lime plaster have been reviewed (54). Carbonate deposition, including both inorganic solutions and biomineralization, barium hydroxide treatment, ammonium oxalate and tartaric acid treatments, have been detailed. These compounds are shown in Figure 5.2.

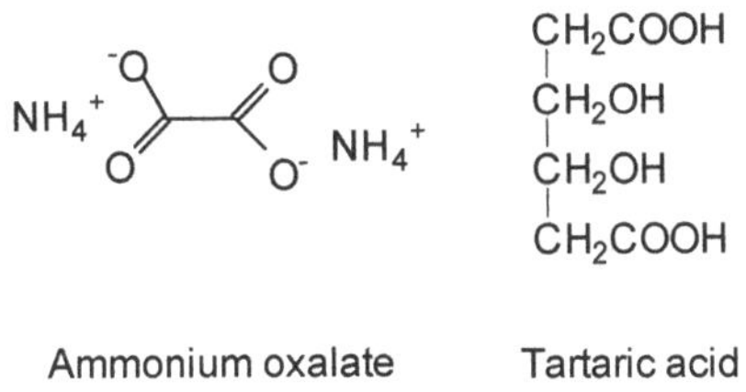

Figure 5.2 Ammonium oxalate and tartaric acid.

5.3.3 *Conservation of Gypsum*

Gypsum-based mortars and plasters have been used since the advent of pyrotechnology, from around 12,000 BC, in architectural applications as joint mortars, to cover masonry or to support paintings and decorations (55–57).

The gypsum technology uses an initial thermal treatment of gypsum at 120–160°C in order to get the hemihydrate:

$$CaSO_4 \times 2H_2O \rightarrow CaSO_4 \times \frac{3}{2}H_2O + \frac{1}{2}H_2O \quad (5.1)$$

When the hemihydrate is mixed with water, the fresh mix rapidly sets and rapidly hardens by the formation of an interconnected porous structure of acicular or prismatic gypsum crystals. These crystals impart the strength to the plasterwork. Therefore, gypsum is suitable for a lot of architectural and decorative applications.

Gypsum plasterworks and decorative surfaces are easily degraded, particularly when exposed to humidity. Thus they often require protection and consolidation (58).

However, the conservation of historical gypsum-based structural and decorative materials by conventional organic and inorganic consolidants has shown only a limited efficiency.

A new method based on the bioconsolidation capacity of carbonatogenic bacteria to inhabit the material has been described (58). The method was exemplified on historical gypsum plasters and also compared with conventional consolidation treatments. Conventional consolidation treatments use ethyl silicate, a methylacrylate ethyl methacrylate copolymer or poly(vinyl butyral).

Conventional products do not reach an in-depth consolidation. They typically form a thin impervious surface layer which blocks pores. In contrast, a bacterial treatment produces a vaterite ($CaCO_3$) biocement, which does not block the pores and produces a good level of consolidation, both on the surface and also in depth. This has been shown by the measurement of the drilling resistance. Transmission electron microscopy analysis indicated that bacterial vaterite cement formed via oriented aggregation of $CaCO_3$ nanoparticles with a size of around 20 *nm*. This process results in mesocrystals, which incorporate the bacterial biopolymers (58).

Such a biocomposite has superior mechanical properties. This explains the fact that the drilling resistance of bioconsolidated gypsum plasters is within the range of inorganic calcite materials of equivalent porosity, despite the fact that the bacterial vaterite cement accounts for only a 0.02 solid volume fraction.

For these reasons, bacterial bioconsolidation has been proposed for the effective consolidation of this type of material.

5.3.4 *Stone Tools*

Stone tools dominate archaeological assemblages from all periods in most areas of the world. Monographs that deal with this issue are available (59,60).

The researchers who analyze prehistoric stone tools and their production debris have made significant progress in understanding the relationship between stone tools and human organizational strategies (61).

Stone tools are understood to be morphologically dynamic throughout their use lives. The changing morphology of stone tools is intimately associated with the needs of the users of the stone tools.

The recent literature on stone tool production has been reviewed with an emphasis on raw material procurement, manufacturing techniques, and tool maintenance processes as they relate to adaptive strategies of the producers of the stone tools and the users of stone tools (61).

5.3.4.1 *Sites and Issues*

The Oldowan stone tool industry was named for 1.8 million year old artefacts found near the bottom of Olduvai Gorge, Tanzania (62). Subsequent archaeological research in Omo (Ethiopia) and Turkana (Kenya) also yielded stone tools dating back 2.3 million years.

Palaeoanthropological investigations in the Hadar region of the Awash Valley of Ethiopia (63), revealed Oldowan assemblages in the adjacent Gona River drainage (64).

Field work was conducted in the Gona study area of Ethiopia between 1992 and 1994, which resulted in additional archaeological discoveries as well as radio-isotopic age control and magnetic polarity stratigraphy of the Gona sequence (62). These occurrences could now be securely dated between 2.6–2.5 million years old.

Thus, these stone tools are the oldest known artefacts from anywhere in the world. The artefacts exhibit a surprisingly sophisticated control of the stone fracture mechanics, equivalent to much younger Oldowan assemblages of the Early Pleistocene age. This indicates an unexpectedly long period of technological stasis in the Oldowan (62).

A Late Prehistoric period buffalo kill and butchering site in northern Wyoming produced a large number of stone tools (65). Flakes removed in sharpening stone tools provided much of the interpretation of the activity that occurred at the site and in addition provided a number of ideas concerning tool use and sharpening.

5.3.5 *Bioreceptivity of Glazed Tiles*

A laboratory-based methodology was developed to assess the bioreceptivity of glazed tiles. The methodology consisted of multiple steps (66):

1. Manufacturing of pristine and artificially aged glazed tiles,
2. Enrichment of phototrophic microorganisms,
3. Inoculation of phototrophs on glazed tiles,
4. Incubation under optimal conditions, and
5. Quantification of the biomass.

Also, the intrinsic properties of the tiles were assessed in order to determine the specific material properties that contribute to the bioreceptivity of the tiles.

The growth rate of the biofilm and the biomass were examined by digital image analysis, colorimetry and chlorophyll A analysis. SEM, micro-Raman and micro-particle induced X-ray emission analysis were carried out to assess the biodeteriorating potential of the phototrophic microorganisms on the glazed tiles (66).

These investigations showed that the accelerated colonization conditions allowed different types of tile bioreceptivity to be distinguished which are associated with the specific characteristics of the particulate material. Aged tiles showed a higher bioreceptivity than pristine tiles due to their higher capillarity and permeability. On colonized tile surfaces a biophysical deterioration caused by a chasmoendolithic growth was observed (66).

5.3.6 *Rock Art Protection*

The design of a rock art protective structure at Petroglyphs Provincial Park, Ontario, Canada, has been described (67). This site was damaged by frost and algal weathering and was vulnerable to vandalism.

A structure was built over the site to control the access of visitors and to protect it from rain and snow in order to mitigate biodeterioration and physical weathering. Earlier studies of the site, following its rediscovery in 1954, have been reviewed. The site was recorded using raking light photography, conventional photography, and stereophotogrammetry. The preservation, interpretive, operational, and design criteria for the structure have been described.

The protective structure was opened in May 1985. The site has been stabilized and further natural weathering has been prevented. The structure provides for controlled visitor access to the site, including access for the disabled, visitor circulation around the site, unobstructed viewing of the petroglyphs, and evening viewing of the petroglyphs with raking light illumination (67).

5.3.7 *Polymers*

5.3.7.1 *Acrylic Copolymers*

Treatments for stone conservation based on acrylic resins have been established as reversible. This occurs due to their solubility in certain solvents (3, 68–70). However, even if a protective treatment is easily soluble, repeated applications of a solvent may cause damage to the material (71).

Acrylic copolymers are now well known in conservation practice. Based upon relative stability and reversibility, Paraloid B-72 has found a particularly wide range of uses as a coating, consolidant, and adhesive (72). Increasingly, Paraloid B-72 has been used as an adhesive for reassembly of large stone sculpture.

It has also found use as an interface barrier that, in theory, lends greater reversibility to joints made with less reversible adhesives, such as epoxies or polyesters, and limits the penetration of these materials into the adherend substrate.

Two studies were performed for an initial evaluation of these systems and to make a direct comparison between Paraloid B-72 and the more commonly used structural adhesives, i.e., epoxy and polyester, from the viewpoint of shear and tensile strength, as well as reversibility.

Modified ASTM standard tests were used to determine the strength of marble-to-marble bonds formed with epoxy, polyester, and high concentration solutions of Paraloid B-72 in two commonly used solvents.

Bonds made with epoxy or polyester adhesives, including additional barriers of Paraloid B-72, on both substrates of the adherends were also tested. The results indicate that in some solutions Paraloid B-72 alone may be sufficiently strong under tensile load for use as a structural adhesive and that Paraloid B-72 used as an interface barrier is sufficiently strong for structural joints.

Also, the reversibility of bonds can be increased where polyester or epoxy is used as adhesive. However, the solvent loss from the Paraloid B-72 adhesive or Paraloid B-72 barrier must be sufficient to guarantee a full bond strength (72).

5.3.7.2 Epoxy-Silica Polymers

The reaction of the epoxy derivatives 2-(3,4-epoxycyclohexyl)ethyl-trimethoxysilane and (3-glycidyloxypropyl)methyldiethoxysilane with the primary amine (3-aminopropyl)triethoxysilane results in solids that have been investigated by means of TG-DTA, FT-Raman and Fourier transform infrared spectroscopy (73). Solution ^{13}C NMR and solid-state FT-Raman spectra show that the oxirane ring of 2-(3,4-epoxycyclohexyl)ethyl-trimethoxysilane doe not take part in the epoxy-amine addition.

The structure of the resulting materials consists of a siloxane network with pendant epoxycyclohexyl groups. In contrast, when (3-glycidyloxypropyl)methyldiethoxysilane is allowed to react with (3-aminopropyl)triethoxysilane, the polymeric network develops either on the organic or inorganic side, so that the resulting materials may be viewed as silica-epoxy hybrids with organic polymer chains covalently linked to inorganic domains. The compounds mentioned above are shown in Figure 5.3.

Porosity, water absorption properties and dynamic contact angle measurements were used for preliminary characterization of the efficacy of the above-described materials as conservation products (73).

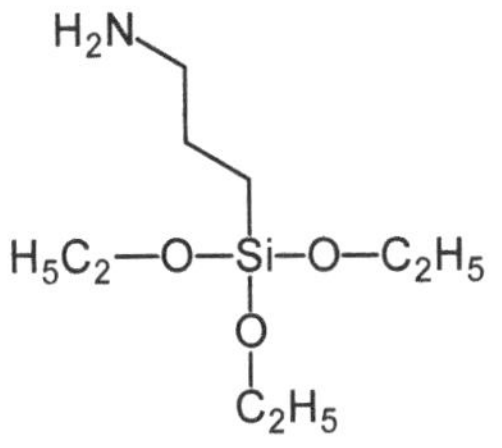

(3-Aminopropyl)triethoxysilane

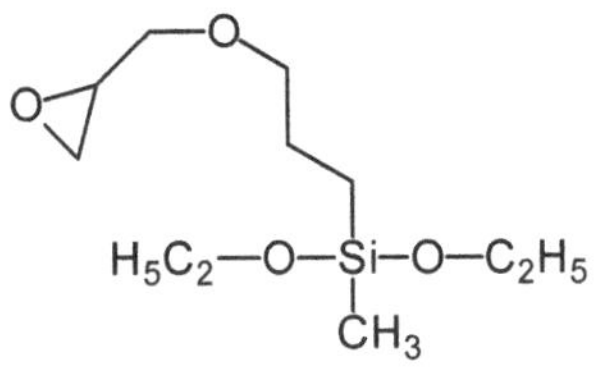

3-Glycidyloxypropyl)methyldiethoxysilane

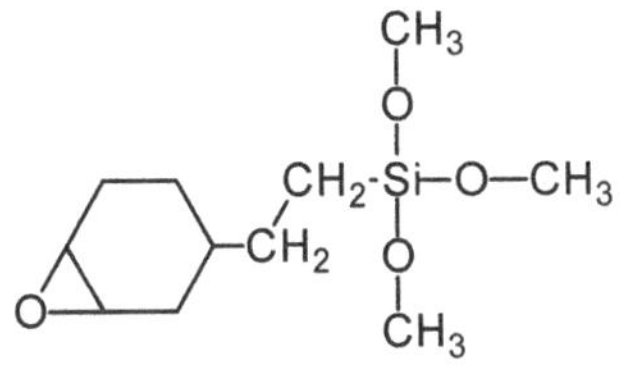

2-(3,4-Epoxycyclohexyl)ethyl-trimethoxysilane

Figure 5.3 Silane compounds.

5.3.8 Biocalcification Treatment

Among the techniques used to limit or stop the penetration of water into the stone, the calcification properties of bacteria have been investigated and used to treat buildings (74).

The effects of such a treatment following a protocol used *in-situ* have been studied. The effects of the biotreatment on limestone (*Tuffeau*) were measured in the course of a large number of drying-imbibition cycles.

The imbibition curves did not follow the usual Washburn law. Therefore, a model based on a space-dependent permeability coefficient has been proposed. This results in a nonlinear diffusion model which accounts for the deviation from the standard Washburn model.

The Washburn equation (75) describes the capillary flow in porous media. This is defined as (76):

$$L^2 = \frac{\gamma D t}{4\varepsilon} \tag{5.2}$$

Here, t is the time that a liquid with a viscosity of ε and a surface tension of γ needs to enter a distance of L into a porous material with an average pore size of D. By the way, the Washburn equation is of importance in a series of other physical fields such as inkjet printing.

5.3.9 *Water Repellent Treatment*

The performance of a water repellent treatment applied to a stone surface is usually evaluated without considering the depth of penetration and uniformity of distribution of the active product in the treated porous material (77).

Magnetic resonance imaging is a nondestructive method. This technique has made it possible to visualize the presence and distribution of different hydrophobic polymers in rocks.

It has been shown, by using this technique, how the penetration depth and the distribution of a water repellent material in a stone are affected. Also, this may to some extent be directed by the method of application. Two commercial hydrophobic polymers, Paraloid B-72 and Silirain® 50, that are frequently used in the restoration and conservation of stone artifacts have been tested on a biocalcarenite (Pietra di Leece) (77).

5.3.10 *Calcium Hydroxide Nanoparticles*

Slaked lime ($Ca(OH)_2$) nanoparticles were exposed at 33% and 75% relative humidity to consolidate dolostone samples used in historical buildings (78).

Nondestructive techniques were applied to determine the chemical, morphological, physical and hydric properties of the stone samples, before and after 20 *d* treatment. Morphological and mineralogical characterization of the nanoparticles was performed.

It has been found that 75% relative humidity favors the consolidation process. This was studied using environmental scanning electron microscopy, spectrophotometry, capillarity, water absorption under vacuum, ultrasound velocity, NMR, both imaging and relaxometry, and optical surface roughness analysis.

At 75% relative humidity, the nanoparticles fill the pores and intercrystalline dolomite grain contacts but do not favor calcite recrystallization, as it occurs at 33% relative humidity. Environmental scanning electron microscopy, XRD, and transmission electron microscopy measurements under 75% relative humidity show a fast transformation of portlandite ($Ca(OH)_2$) into vaterite ($CaCO_3$), monohydrocalcite ($CaCO_3 \times H_2O$) and calcite ($CaCO_3$). This eventually improves the physical and hydric properties of the stones significantly.

The propagation of the ultrasound velocity was measured before and after 20 *d* of the treatment to evaluate the increase in durability and the distribution of the consolidating product, for the P-wave velocity, which is related to the effective porosity. P-waves are known in seismology. These waves travel through a continuum and are the first waves that arrive from an earthquake. The term *P-wave* stands for either pressure wave or primary wave.

Water absorption under vacuum conditions was performed to determine the bulk density, open porosity and also to quantify the amount of water absorbed by the specimens once they reach saturation.

The NMR was measured before and after a 20 *d* treatment in order to observe and to quantify the location and distribution of water and consolidating product inside the pores of the stone specimens.

Thus, new insights have been provided for the assessment of consolidation effectiveness of porous carbonate stones with calcium hydroxide nanoparticles under optimum relative humidity conditions combining several nondestructive techniques (78).

5.3.11 *Nanolime Calcium Hydroxide with Triton*

Calcium carbonate is compatible with many carbonate-based architectonic surfaces, because its characteristics are similar to those of the restored materials. In order to improve calcium hydroxide treatments, $Ca(OH)_2$ particles with submicrometric dimensions have been synthesized by a chemical precipitation process (79): A sodium hydroxide solution, used as precipitator, is added, drop by drop, into a calcium chloride solution.

A nanometric calcium hydroxide, to be used in stone treatment, has been produced by adding a surfactant agent (Triton X-100) into the initial solutions. The solutions are then mixed together simultaneously, thus drastically reducing the time needed for preparation. Different contents of surfactant are employed, and the influence on particle dimension and carbonatation process has been analyzed.

The obtained $Ca(OH)_2$ nanoparticles have been characterized by XRD, SEM and by electron diffraction. The $Ca(OH)_2$ nanoparticles are regularly shaped, hexagonally plated and with side dimensions less than 200 *nm*. By increasing the surfactant content, the particle dimension reaches values up to 20 *nm*.

Comparing nanolime suspensions obtained without and with the addition of the surfactant agent, the latter have comparable, if not better, performances in terms of average particle size and morphology, crystallinity and reactivity.

The alcoholic nanolime suspension with Triton X-100 has been applied on some natural stones (79). Triton X-100 is a poly(ethylene glycol) *tert*-octylphenyl ether with 9 to 10 ethylene oxide side groups. It is a widely used nonionic surfactant with mild non-denaturing conditions.

5.3.12 *Nanocomposites for the Protection of Granitic Obelisks*

Granite has been widely used in the fields of architecture and sculpture in Egypt, especially in the carving of obelisks (80). Many granitic obelisks of Ramses II were found in the ancient city of Tanis (San El Hagar), in the Nile Delta of Egypt. This city has been considered one of the most important ancient Egyptian cities.

Unfortunately, these granitic obelisks have been subjected to many deterioration processes. This has resulted in various deterioration issues, such as (80):

- Granular disintegration,
- Scaling or spalling,
- Cracking,
- Efflorescence,
- Soiling, and
- Microbiological colonization.

Thus, the materials which are used in the protection procedures of such granitic obelisks must have the property of hydrophobicity in order to protect them from the harmful effects of water.

Recently, polymeric nanoparticle composites have attracted great interest in the field of cultural heritage conservation due to their unique multifunctional properties. Important issues are the high surface area and the chemical activity of the nanoparticles that are dispersed in the polymers.

Three types of nanocomposites, i.e., PF 4, Fluotanium, and Fluozinc, were used for the protection of granitic samples collected from Tanis (80).

PF 4 is a ready-to-use product, based on titanium dioxide nanoparticles in an alkoxy silane polymer (81).

The other two nanocomposites can be prepared by adding titanium dioxide and zinc oxide nanoparticles to a fluorocarbon commercial polymer, such as PK 50 AE.

The results of petrographical, mineralogical, and chemical characterization indicted that the granitic obelisks were subjected to hard weathering, which resulted in a partial alteration of both biotite and potash feldspar into sericite. This arises due to the leaching of iron from the crystal structure of biotite, oxidation of ferrous ions into ferric ions thus forming iron oxide stains on microfracture cleavages.

All nanocomposites used in the study enhanced the property of water repellency of the granitic samples. In particular, PK 50 AE reached high degrees of contact angle. Furthermore, the nanocomposites reduced the amount of water absorption. The decrease in water absorption was found to be in the range of 88–93%.

Also, the treatments resulted in a slight improvement of the abrasion resistance of the materials. This can be attributed to the low amount of polymer uptakes resulting from the low porosity of the granitic samples. The polymer uptake in the treated granitic samples was found to be 0.04–0.05% (80).

5.3.13 *Superhydrophobic Films*

A simple method has been described that can be used to impart superhydrophobicity to stone surfaces of monuments using common and low-cost materials (82).

A siloxane-nanoparticle dispersion is sprayed on a stone, and this process can result in the formation of a rough two-length-scale hierarchical structure that exhibits water repellent properties, provided that the nanoparticle concentration in the dispersion is higher than a critical value. Superhydrophobicity (static contact angle >150° and contact angle hysteresis <7°) is achieved by this simple method (82):

1. On the surfaces of three types of stones, Opuka, Božanovský and Horický, which have been used for the restoration of the castle of Prague,
2. Using two poly(alkyl siloxane) products such as Rhodorsil® 224 and Porosil® VV plus, which are utilized by conservators, and
3. Using common nanoparticles such as silica (SiO_2), alumina (Al_2O_3), tin oxide (SnO_2) and titanium oxide (TiO_2).

It has been shown that the stone substrate and the nanoparticle size (5–50 *nm*) or type have almost no effect on the wettability of the superhydrophobic surfaces, as comparable contact angles were measured on the three stone substrates, treated with any siloxane-particle composite.

Treatments of the stones with pure hydrophobic siloxanes and superhydrophobic siloxane silica composites result in comparable reductions of the water vapor permeability and the water amounts absorbed by capillarity.

Thus, the use of nanoparticles in protective coatings does not have any obvious effect on the results of the aforementioned procedures. However, the aesthetic appearance of the three stones included in this study is highly affected by the nanoparticles (82).

References

1. M. Forsyth, *Interior Finishes & Fittings for Historic Building Conservation*, Wiley-Blackwell, Chichester, West Sussex, UK, Ames, Iowa, 2012.
2. D. Watt, *Surveying Historic Buildings*, Routledge, New York, 2nd edition, 2015.
3. C.A. Price and E. Doehne, *Stone Conservation: An Overview of Current Research*, Research in Conservation, Getty Publications, Los Angeles, 2nd edition, 2010.
4. C. Pearson, *Conservation of ceramics, glass and stone* in C. Pearson, ed., *Conservation of Marine Archaeological Objects*, chapter 11, pp. 253–267. Butterworths-Heinemann, London, 2014.
5. A. Henry, ed., *Stone Conservation: Principles and Practice*, Routledge, New York, 2015.

6. C. Pearson, *Deterioration of ceramics, glass and stone* in C. Pearson, ed., *Conservation of Marine Archaeological Objects*, chapter 11, pp. 99–104. Butterworths-Heinemann, London, 2014.
7. C. Ferrari, G. Santunione, A. Libbra, A. Muscio, E. Sgarbi, C. Siligardi, and G.S. Barozzi, *Int. J. DNE*, Vol. 10, p. 21, March 2015.
8. T. Warscheid and J. Braams, *International Biodeterioration & Biodegradation*, Vol. 46, p. 343, 2000.
9. J.W. Hopton, *Physical conditions and microbial growth: Some implications for biodeterioration* in D.R. Houghton, R.N. Smith, and H.O.W. Eggins, eds., *Biodeterioration*, chapter 7, pp. 511–516. Elsevier Applied Science, London, New York, 1988.
10. R.J. Palmer and P. Hirsch, *Geomicrobiology Journal*, Vol. 9, p. 103, January 1991.
11. E. Zanardini, P. Abbruscato, N. Ghedini, M. Realini, and C. Sorlini, *International Biodeterioration & Biodegradation*, Vol. 45, p. 35, January 2000.
12. M.L. Coutinho, A.Z. Miller, and M.F. Macedo, *Journal of Cultural Heritage*, Vol. 16, p. 759, 2015.
13. A.G. Nord and T. Ericsson, *Studies in Conservation*, Vol. 38, p. 25, January 1993.
14. F.E.W. Eckhardt, *Solubilization, transport, and deposition of mineral cations by microorganisms-efficient rock weathering agents* in J. Drever, ed., *The Chemistry of Weathering*, pp. 161–173. D. Reidel Publ. Comp. Ltd., Dordrecht, 1985.
15. M.A. de la Torre, G. Gomez-Alarcon, C. Vizcaino, and M.T. Garcia, *Biogeochemistry*, Vol. 19, p. 129, January 1992.
16. F. Jroundi, M.T. Gonzalez-Muñoz, C. Rodriguez-Navarro, B. Martin-Peinado, and J. Martin-Peinado, Conservation of carbonate stone by means of bacterial carbonatogenesis: Evaluation of in situ treatments, in *Proceedings of the 8th International Symposium on the Conservation of Monuments in the Mediterranean Basin. Patras: Technical Chamber of Greece*, Vol. 3, pp. 159–171, 2013.
17. M.T. Madigan, J.M. Martinko, and J. Parker, *Biology of Microorganisms*, Vol. 514, Prentice Hall, Upper Saddle River, NJ, 1997.
18. K. Milde, W. Sand, W. Wolff, and E. Bock, *Microbiology*, Vol. 129, p. 1327, May 1983.
19. E. Bock and W. Sand, *Journal of Applied Bacteriology*, Vol. 74, p. 503, 1993.
20. A.A. Gorbushina, W.E. Krumbein, C.H. Hamman, L. Panina, S. Soukharjevski, and U. Wollenzien, *Geomicrobiology Journal*, Vol. 11, p. 205, 1993.
21. E. May, F.J. Lewis, S. Pereira, S. Tayler, M.R.D. Seaward, and D. Allsopp, Microbial deterioration of building stone: A review, in *Biodeterioration Abstracts*, Vol. 7, pp. 109–123, 1993.

22. T. Warscheid, T.W. Becker, J. Braams, C. Gehrmann, W.E. Krumbein, K. Petersen, and S. Bruggerhoff, Studies on the temporal development of microbial infection of different types of sedimentary rocks and its effect on the alteration of the physico-chemical properties in building materials, in *Conservation of Stone and Other Materials*, Vol. 1, pp. 303–310, Paris, 1993. Proceedings of the International RILEM/UNESCO Congress, E&F N. Spon.
23. C.A. Crispim and C.C. Gaylarde, *Microbial Ecology*, Vol. 49, p. 1, 2005.
24. M. Wagner and W. Schwartz, *Zeitschrift für allgemeine Mikrobiologie*, Vol. 7, p. 33, 1967.
25. E. Bock, B. Ahlers, and C. Meyer, *Bauphysik*, Vol. 11, p. 141, 1989.
26. P.M. Gaylarde and C.C. Gaylarde, *International Biodeterioration & Biodegradation*, Vol. 46, p. 93, September 2000.
27. B.O. Ortega-Morales, A. López-Cortés, G. Hernandez-Duque, P. Crassous, and J. Guezennec, *Extracellular polymers of microbial communities colonizing ancient limestone monuments* in R.J. Doyle, ed., *Microbial Growth in Biofilms - Part A: Developmental and Molecular Biological Aspects*, Vol. 336 of *Methods in Enzymology*, pp. 331–339. Academic Press, San Diego, 2001.
28. A.A.O.D. El Derby, M.M.A. Mansour, and M.Z.M. Salem, *Mediterranean Archaeology and Archaeometry*, Vol. 16, p. 273, 2016.
29. D.R. Piperno and I. Holst, *Journal of Archaeological Science*, Vol. 25, p. 765, August 1998.
30. Y. Yang, W. Li, L. Yao, Z. Cheng, W. Luo, J. Zhang, L. Lin, H. Gan, and L. Yan, *Science China Earth Sciences*, Vol. 59, p. 1574, 2016.
31. A. Pedergnana, L. Asryan, J.L. Fernández-Marchena, and A. Ollé, *Micron*, Vol. 86, p. 1, July 2016.
32. V. Rots, B.L. Hardy, J. Serangeli, and N.J. Conard, *Journal of Human Evolution*, Vol. 89, p. 298, December 2015.
33. G.H. Langejans, *Journal of Archaeological Science*, Vol. 37, p. 971, 2010.
34. T.H. Loy, *Methods in the analysis of starch residues on prehistoric stone tools* in J.G. Hather, ed., *Tropical Archaeobotany: Applications and New Developments*, chapter 5, pp. 86–114. Routledge, London, New York, 1994.
35. L.C. Prinsloo, L. Wadley, and M. Lombard, *Journal of Archaeological Science*, Vol. 41, p. 732, January 2014.
36. I. Liritzis and N. Droseros, *Mediterranean Archaeology and Archaeometry*, Vol. 15, p. 277, 2015.
37. D.J. Huntley, D.I. Godfrey-Smith, and M.L.W. Thewalt, *Nature*, Vol. 313, p. 105, January 1985.
38. I. Liritzis, A.K. Singhvi, J.K. Feathers, G.A. Wagner, A. Kadereit, N. Zacharias, and S.-H. Li, *Luminescence Dating in Archaeology, Anthropology, and Geoarchaeology: An Overview*, Springer Briefs in Earth System Sciences, Springer, Cham, New York, 2013.

39. X. Miao, H. Wang, P.R. Hanson, J.A. Mason, and X. Liu, *Geoderma*, Vol. 261, p. 93, 2016.
40. M. Camaiti, V. Bortolotti, and P. Fantazzini, *Magnetic Resonance in Chemistry*, Vol. 53, p. 34, 2015.
41. M. Brai, M. Camaiti, C. Casieri, F. De Luca, and P. Fantazzini, *Magnetic Resonance Imaging*, Vol. 25, p. 461, May 2007.
42. R.J.S. Brown and P. Fantazzini, *Phys. Rev. B*, Vol. 47, p. 14823, June 1993.
43. E.L. Hahn, *Physical Review*, Vol. 80, p. 580, November 1950.
44. Wikipedia, Spin echo — wikipedia, the free encyclopedia, 2016. [Online; accessed 10-August-2016].
45. B. Blümich, F. Casanova, J. Perlo, S. Anferova, V. Anferov, K. Kremer, N. Goga, K. Kupferschläger, and M. Adams, *Magnetic Resonance Imaging*, Vol. 23, p. 197, February 2005.
46. G.C. Borgia, M. Camaiti, F. Cerri, P. Fantazzini, and F. Piacenti, *Journal of Cultural Heritage*, Vol. 1, p. 127, June 2000.
47. L. Appolonia, G.C. Borgia, V. Bortolotti, R.J. Brown, P. Fantazzini, and G. Rezzaro, *Magnetic Resonance Imaging*, Vol. 19, p. 509, April 2001.
48. G.C. Borgia, V. Bortolotti, M. Camaiti, F. Cerri, P. Fantazzini, and F. Piacenti, *Magnetic Resonance Imaging*, Vol. 19, p. 513, April 2001.
49. M. Alesiani, S. Capuani, B. Maraviglia, R. Giorgi, and P. Baglioni, *Applied Magnetic Resonance*, Vol. 23, p. 63, October 2002.
50. M. Camaiti, C. Casieri, F. De Luca, P. Fantazzini, and C. Terenzi, *Studies in Conservation*, Vol. 52, p. 37, 2007.
51. M. Brai, C. Casieri, F. De Luca, P. Fantazzini, M. Gombia, and C. Terenzi, *Solid State Nuclear Magnetic Resonance*, Vol. 32, p. 129, December 2007.
52. C. Alves and J. Sanjurjo-Sánchez, *Environ. Chem. Lett.*, Vol. 13, p. 413, August 2015.
53. N. Perez-Ema, M. Alvarez de Buergo, R. Bustamante, and M. Gomez-Heras, *Changes in petrophysical properties of the stone surface due to past conservation treatments in archaeological sites of Merida (Spain)* in G. Lollino, D. Giordan, C. Marunteanu, B. Christaras, I. Yoshinori, and C. Margottini, eds., *Engineering Geology for Society and Territory - Volume 8: Preservation of Cultural Heritage*, pp. 521–524. Springer International Publishing, Cham, 2015.
54. E. Hansen, E. Doehne, J. Fidler, J. Larson, B. Martin, M. Matteini, C. Rodriguez-Navarro, E.S. Pardo, C. Price, A. de Tagle, J.M. Teutonico, and N. Weiss, *Studies in Conservation*, Vol. 48, p. 13, 2003.
55. J. Ashurst and N. Ashurst, *Practical Building Conservation: Mortars, Plasters and Renders*, Vol. 3 of *English Heritage Technical Handbook*, Gower Technical Press, Aldershot, UK, 1988.
56. W.D. Kingery, P.B. Vandiver, and M. Prickett, *Journal of Field Archaeology*, Vol. 15, p. 219, 1988.

57. J. Elsen, *Cement and Concrete Research*, Vol. 36, p. 1416, August 2006.
58. F. Jroundi, M.T. Gonzalez-Muñoz, A. Garcia-Bueno, and C. Rodriguez-Navarro, *Acta Biomaterialia*, Vol. 10, p. 3844, September 2014.
59. G. Odell, ed., *Stone Tools: Theoretical Insights into Human Prehistory*, Springer Science+Business Media, LLC, New York, 1996.
60. M.R. Edmonds, *Stone Tools and Society: Working Stone in Neolithic and Bronze Age Britain*, Routledge, London, 2001.
61. W. Andrefsky, *Journal of Archaeological Research*, Vol. 17, p. 65, 2009.
62. S. Semaw, P. Renne, J.W.K. Harris, C.S. Feibel, R.L. Bernor, N. Fesseha, and K. Mowbray, *Nature*, Vol. 385, p. 333, January 1997.
63. D.C. Johanson, M. Taieb, and Y. Coppens, *American Journal of Physical Anthropology*, Vol. 57, p. 373, 1982.
64. G. Corvinus, *Nature*, Vol. 261, p. 571, June 1976.
65. G.C. Frison, *American Antiquity*, Vol. 33, p. 149, April 1968.
66. M.L. Coutinho, A.Z. Miller, M.A. Rogerio-Candelera, J. Mirao, L.C. Alves, J.P. Veiga, H. Águas, S. Pereira, A. Lyubchyk, and M.F. Macedo, *Biofouling*, Vol. 32, p. 243, 2016. PMID: 26900634.
67. I.N.M. Wainwright, H. Sears, and S. Michalski, *Journal of the Canadian Association for Conservation*, Vol. 22, p. 53, 1997.
68. D. Erhardt, *Journal of the American Institute for Conservation*, Vol. 22, p. 100, January 1983.
69. A.E. Charola, A. Tucci, and R.J. Koestler, *Journal of the American Institute for Conservation*, Vol. 25, p. 83, January 1986.
70. B. Appelbaum, *Journal of the American Institute for Conservation*, Vol. 26, p. 65, January 1987.
71. M. Gómez-Heras, M. Alvarez de Buergo, E. Rebollar, M. Oujja, M. Castillejo, and R. Fort, *Applied Surface Science*, Vol. 219, p. 290, 2003.
72. J. Podany, K.M. Garland, W.R. Freeman, and J. Rogers, *Journal of the American Institute for Conservation*, Vol. 40, p. 15, Spring 2001.
73. P. Cardiano, R. Ponterio, S. Sergi, S.L. Schiavo, and P. Piraino, *Polymer*, Vol. 46, p. 1857, 2005.
74. O. Rozenbaum, S. Anne, and J.-L. Rouet, *Construction and Building Materials*, Vol. 70, p. 97, November 2014.
75. E.W. Washburn, *Physical Review*, Vol. 17, p. 273, March 1921.
76. Wikipedia, Washburn's equation — wikipedia, the free encyclopedia, 2016. [Online; accessed 25-November-2016].
77. G.C. Borgia, M. Camaiti, F. Cerri, P. Fantazzini, and F. Piacenti, *Studies in Conservation*, Vol. 48, p. 217, 2003.
78. P. López-Arce, L.S. Gomez-Villalba, L. Pinho, M.E. Fernández-Valle, M.l. de Buergo, and R. Fort, *Materials Characterization*, Vol. 61, p. 168, February 2010.
79. V. Daniele and G. Taglieri, *Journal of Cultural Heritage*, Vol. 13, p. 40, 2012.

80. F.M. Helmi and Y.K. Hefni, *Mediterranean Archaeology and Archaeometry*, Vol. 16, p. 87, 2016.
81. CHEM SPEC Innovative Chemical Distribution, Innovative distribution chemicals, electronic: , 2016.
82. P.N. Manoudis, I. Karapanagiotis, A. Tsakalof, I. Zuburtikudis, B. Kolinkeová, and C. Panayiotou, *Applied Physics A*, Vol. 97, p. 351, April 2009.

6

Glass

There are monographs concerning the conservation and restoration of glass (1–5). Also, the importance of corrosion of glass and the issues and approaches for the archaeological science have been described (6).

6.1 Analytical Methods

6.1.1 Spectrometric Investigation of Weathering Processes

Nowadays, atmospheric pollution plays an important role in the weathering of the historical buildings and glass windows (7). Samples of white powdered weathering products, recovered during the restoration of the stained glass windows of León Cathedral in Spain, were collected. The white powder was found to be distributed on the glass in small outbreaks adjacent to the lead frame, directly deposited on the lead, and also on different areas of the glass surfaces.

The samples were characterized using a combination of scanning electron microscope (SEM) with energy dispersive X-ray spectroscopy (EDX), Fourier transform infrared (FTIR) spectroscopy and Raman spectrometry. The presence of sulfates and smaller amounts of carbonates in the white powdered product is a clear indication of the participation of atmospheric acidifying gases, in particular SO_X, in the course of the weathering process.

The following reactions refer to transformations from carbonates and carbonaceous materials, particularly in a humid environment (7):

$$CaC_2O_4 + \frac{1}{2}O_2 \rightarrow CaCO_3 + CO_2 \tag{6.1}$$

$$CaCO_3 + SO_2 + \frac{1}{2}O_2 \rightarrow CaSO_4 \times H_2O + H_2CO_3 \tag{6.2}$$

$$CaO + SO_3 \rightarrow CaSO_4 \tag{6.3}$$

Also, reaction of SO_2 in the presence of alkali elements may effect the deterioration of a glass surface (7,8):

$$3SO_2 + 2Na_2O \rightarrow 2Na_2SO_4 + S \tag{6.4}$$

However, the experiments did not indicate the participation of NO_x gases. On the other hand, there was some evidence that the putty and mortar used to seal or join the glasses were major sources of the weathering products.

As a result, it was suggested to use sealants that are more resistant to oxidation, such as silicone-based and zirconia-based materials for repairing glass windows in historic buildings in order to avoid exacerbating degradation (7).

6.1.2 *Analysis of Historic Glass*

In archaeometric studies, glass materials are usually studied by X-ray fluorescence spectroscopy or SEM-EDX (9). It has also been demonstrated that optical spectroscopy can be used in a straightforward way to identify the band positions of the coloring materials. Cobalt is an efficient marker to identify the fluxes used for glass production. Therefore, this analysis can help to distinguish medieval and post-medieval materials from pieces added in the course of more recent restoration campaigns.

Roman, post-medieval, and industrial glasses have been analyzed by optical spectroscopy. There, two compositional groups could be identified: A calco-potassic group and a sodic group. It was shown that in calco-potassic glasses cobalt showed a shift towards shorter wavelengths with respect to the cobalt absorption bands in soda glass (9).

6.1.3 *Optical Spectroscopy*

Optical spectroscopy has been tested for a first-line analysis of green colored glass artefacts from the Roman period. It has been demonstrated that the color coordinates calculated from the measured transmission spectrum could give information about the fragment under study. In particular it has been shown that (10):

1. Based on the position of the calculated color values on the color diagram one could easily determine whether the artefact was colored using only iron or if copper oxides were also present. In the case of the artefact owing its color solely to the presence of iron, the distance between the measured color values and the white point in the color diagram can roughly indicate the iron concentration of the sample,
2. Artefacts that were fabricated under similar furnace conditions can also be identified on the color diagram, and
3. Samples with identical compositions and fabrication conditions but with different sample thickness, give rise to a variation in the color coordinates, thus allowing optical spectroscopy to help identify fragments which might belong to the same object.

6.1.4 *Portable Raman Spectroscopy*

The advantages of using Raman spectroscopy and X-ray fluorescence spectroscopy for the analysis of glyptics have been discussed (11).

Glyptics are engraved gemstones or glass objects. They sometimes consist of different layered materials. The word glyptic is derived from the Greek word *γλiππω*, a verb indicating engraving or carving hard stones. Engraved gemstones or glass materials were originally used as stamps to identify the owner, for instance, on letters, but also on wine vessels.

A set of 64 glyptics (42 Roman glass specimens and 22 modern ones), belonging to the collection of the museum Quinta das Cruzesâ in Funchal (Madeira, Portugal), was analyzed using portable Raman spectroscopy and handheld X-ray fluorescence. These techniques also can be used to confirm the gemological identification of these precious objects and can give extra information about the glass composition.

With Raman spectroscopy the molecular composition can be identified, as well the crystalline phases present in the material. In addition, handheld X-ray fluorescence indicated that the antique Roman glass samples have low Pb and Sn levels, Thus the modern specimens can be discriminated into two groups: lead-based and non-lead-based ones (11).

6.1.5 3D Laser Ablation Mass Spectrometry

A multi-elemental 3D laser ablation inductively coupled plasma mass spectrometry mapping procedure for high-resolution depth information retrieval to investigate surface layer phenomena has been described (12).

The procedure is based on laser drilling on a virtual grid on the surface, followed by extraction of depth maps along the z-axis for each element monitored. Using a burst of 50 laser pulses at 1 *Hz* on each point of the grid, a penetration rate of around 150 *nm* per pulse was obtained with a lateral resolution in the order of the laser beam diameter.

Using an ultrafast monitoring technique of the individual ablation pulses (58 *ms* for a set of 19 elements) with a laser ablation cell with fast signal washout of less than 0.5 *s* for the whole laser pulse, the corresponding peak areas could be consistently integrated, resulting in spatial elemental data associated with individual pulses.

The usual laser drilling limitations, such as pulse mixing and signal tailing, could be avoided with this approach. After manipulation of the spatial elemental datasets and quantification, stacks of 50 2D depth maps for each element monitored were produced, which could be visualized as volume images or time-lapse movies.

This method was successfully used to investigate the degradation mechanisms of a medieval, weathered glass artifact by colocalization analysis of selected cross-sectional two-dimensional elemental images in arbitrary planes of the volume images.

The results revealed that degradation must have started as a result of dealkalinization, leading to depletion of alkalis/earth alkalis in glass surface layers and an enrichment of the network forming elements Si and Al (12).

6.2 Cleaning Methods

6.2.1 *Medieval Stained Glass Corrosion*

A study on the chemistry of weathering and corrosion of stained glass has been presented (13). Protective methods have been compared and the methods of repair were discussed.

6.2.2 *Effect of Soil pH on the Degradation*

The action of three natural soils types on the degradation of silica glass has been elucidated: acidic, neutral, and basic (14). The mechanisms of corrosion that act on the glasses have been assessed using accelerated aging tests.

Four different glass types were used for the characterization. The chemical composition of these silicate glasses is shown in Table 6.1.

Table 6.1 Chemical composition of the glasses (14).

Glass	Na_2O	MgO	Al_2O_3	SiO_2	P_2O_5	K_2O	CaO
Roman	18.1	1.3	3.4	67.1	-	1.3	7.3
Modern	15.7	1.1	2.3	71.6	-	0.8	8.5
Medieval	1.1	2.5	3.2	47.8	3.4	19.2	22.6
Crystal	5.0	-	0.1	53.8	-	8.2	0.1
Glass	**TiO_2**	**MnO**	**Fe_2O3**	**ZrO_2**	**Sb_2O_3**	**BaO**	**PbO**
Roman	0.1	0.5	0.9	-	-	-	-
Modern	0.1	-	0.1	-	-	-	-
Medieval	-	-	0.1	-	-	0.1	-
Crystal	-	-	-	0.1	0.6	2.6	29.6

The Roman type is a soda lime silicate glass containing a relatively low concentration of Fe_2O_3, to reproduce the impurities present in natural sands, and a low content of MnO to compensate for the color of iron ions. The modern conventional window type is another soda lime silicate glass like the Roman type but with a higher content of both the network formers SiO_2 and CaO. It is colorless due to the very low content of Fe_2O_3. The Medieval type is a potash lime silicate glass. The outstanding difference of this glass in comparison with the two previously mentioned soda lime glasses is a much higher concentration of K_2O and CaO. The lead crystal glass type is a colorless silicate glass, with a high concentration of PbO.

In order to study the behavior of silicate glasses in response to a buried environment, three different accelerated aging burial tests were simultaneously carried out using acidic, neutral, and basic natural soils taken from different places and characterized according to recommended standards set out by the Spanish Society of Soil Science. Thus, the following soil parameters were characterized: textural fraction, pH, content of calcium carbonate and organic matter, and electrical conductivity. To accelerate the aging process, the burial tests were undertaken inside a stove at a constant temperature of 60°C.

The results indicated a distinct and recognizable response of each glass type, with pH as the key variable for glass corrosion.

For example, the Medieval type glass exhibited a different degradation mechanism during the burial tests in comparison with the rest of the silicate glasses. A multilayered structure was formed in which the outermost layers appeared as cracked. In the case of the silicate soda lime glass and lead crystal glasses, the formation of fissures and pits was found, whereas thick degradation layers, often forming multilayered structures, originated in the potash lime glass type. The results have been detailed elsewhere (14).

Such results may have great significance concerning historical glass studies, especially for the selection of suitable conservation strategies for archaeological findings (14).

6.2.3 *Adhesives and Consolidants*

A review of adhesives and consolidants used on glass antiquities has been presented (15).

6.2.4 *Biocorrosion and Biodeterioration*

Various antique and medieval glasses have been examined with regard to general biogenic damage, biopitting, i.e., crater erosion, biocrusts, and opalescent and white biogenic films.

The experiments were carried out using samples that are summarized in Table 6.2.

Table 6.2 Glass samples for evaluation of biocorrosion and biodeterioration (16).

Sample	Origin
Roman glass bottles	Abu Tor, Sinai
Green and blue glass	Cologne Cathedral
Pieces	Church in Évreux
Samples	Fortress of the former Dukedom of Delmenhorst near Oldenburg
Neolithic flint tools	Negev Desert, Israel
Modern glass	Pigsty (19th century)

Some of the bacteria used in these experiments were isolated from the ancient pieces of glass. The biopitting with structures very similar to the biopitting of marble and limestone was found on almost all of the specimens.

Lichens could not be identified directly, but fungi and algae were observed in the pits as well as under the thin layers exfoliated from the Roman glass bottles. The initial steps of colonization and the potential for heavy-metal accumulation due to the isolated bacteria have been shown in laboratory experiments. A fractal dimension of diffusion-limited disaggregation has been suggested as a possible explanation for the characteristic form and the structure of the microbially induced and shaped biopitting patterns. In addition, a biopitting classification has been suggested (16).

6.2.4.1 *Glass Reproductions for Testing Biodeterioration*

Using glass reproductions of historical stained glass windows, their biodeterioration was studied. The chemical composition of the actual historical stained glass windows was determined by micro-energy dispersive X-ray fluorescence. Three special potash glass compositions with different colors were chosen and reproduced (17):

1. A mixed-alkali colorless glass,
2. A purple potash glass with manganese as chromophore, and
3. A brown potash glass colored by iron ions.

The reproduced glass samples were made with two initial surface morphologies, i.e., corroded and non-corroded. The samples were inoculated with fungi previously isolated and identified on the original stained glass windows as species of the genera *Penicillium* and *Cladosporium*.

Both physical and chemical glass surface alterations were analyzed using optical microscopy, Raman microscopy, micro infrared spectroscopy, and scanning electron microscopy with energy dispersive spectroscopy analysis.

The results indicated that the fungi produced a clear damage on all the glass surfaces. The damages were present as spots and stains, fingerprints, biopitting, leaching and the deposition of elements, as well as the formation of biogenic crystals.

Thus it was concluded that the inoculated fungi were able to biodeteriorate glasses with distinct compositions. With regard to the biodeterioration degree, no differences between the initial non-corroded and corroded glass surfaces were observed (17).

6.2.4.2 *Lichens*

A lichen is a composite organism that arises from algae and/or cyanobacteria living among filaments of a fungus in a symbiotic relationship (18)

One of the major roles lichens play in shaping the natural world, both physically and biologically, is as an agent in soil development. Lichens as agents of biodeterioration have been investigated (19). A chemical alteration of the substratum is achieved by the disruptive action of many species, particularly those capable of producing an oxalate at the thallus-substratum interface. The oxalate contributes significantly to the bulk and composition of the thallus itself and persists as an obvious encrustation after the death of the lichen.

6.2.5 *Potash-Lime-Silica Glass*

Glass made from a potash-lime-silica composition was typically produced for cathedral windows during the Middle Ages in Central and Northern Europe (20).

This glass type is known to be very sensitive to the attack of atmospheric pollutants. Potash-lime-silica glass, containing different amounts of K_2O varying from 15% to 25%, was reproduced and aged in concentrated boiling sulfuric acid and also in high temperature-high pressure water, in order to separately simulate the two alteration processes of leaching, i.e., ion exchange and dissolution.

The aged samples were studied by micro-Raman spectroscopy by X-ray diffraction and by SEM EDX technique for the characterization of the newly formed crystalline phases. In this way, a deeper insight into the degradation and stability of these systems could be obtained (20).

The behavior of these glasses as seen by Raman spectroscopy depends on the content of K_2O. Gypsum ($CaSO_4 \times 2H_2O$), bassanite ($CaSO_4 \times 0.5H_2O$) and anhydrite ($CaSO_4$) are found on the alteration layer of the leached samples with higher content of K_2O. In contrast, samples with a content of K_2O of 15% and 20% indicate the presence of the Raman features of the phyllosilicates gyrolite ($Ca_{16}Si_{24}O_{60}(OH)_8 \times 14H_2O$) and reyerite ($KCa_{14}Si_{24}O_{60}(OH)_{\times}5H_2O$) together with hydroxyapatite ($Ca_5(PO_4)_3OH$). In the alteration layer of the glass containing 25% K_2O, the Raman spectra suggest a mixture of charoite ($K_5Ca_8(Si_6O_{15})_2Si_2O_9(OH) \times 3(H_2O)$) and hydroxyapatite (20).

6.2.6 *Chemical Cleaning of Glass*

A chemical strategy has been developed for cleaning soiled deposits and encrustations on archaeological glasses (21).

A series of Roman glass samples, both fragments and complete objects, originating from different excavations in northern Jordan have been tested. The chemical composition of the glass samples was determined by X-ray fluorescence. Further, the mineralogical and elemental composition of the soiled deposits and encrustations on the glass surfaces was investigated by an EDX method. In addition, SEM examination and an optical assessment were performed before and after cleaning glass.

The glass samples were cleaned with different chemicals, such as Calgon, i.e., sodium hexametaphosphate, ethylenediamine tetraacetic acid (EDTA) at different pH values, citric and tartaric acids and piranha solution. Piranha is a solution of sulfuric acid and hydrogen peroxide.

Sepiolite poultices soaked by chemical agents were the most suitable methods used for applying chemical solutions on the glass surface. Also, EDTA is the most effective chelating agent for cleaning encrustations on durable glass. The cleaning was particularly effective and safe at neutral pH at low concentrations of around 5 to 7%. Calcareous crusts could safely be removed by using a piranha solution. Citric and tartaric acids showed only a moderate efficiency on cleaning weathered and stable glass. Calgon has a tendency to damage corroded and iridescent surfaces, and should be avoided when cleaning weathered glass (21).

6.2.7 *Unstable Historic Glass*

A lot of glass objects in museum collections are known to be actively deteriorating, irrespective of the storage or display conditions. Various terms have been used to describe the effect, including sick glass, glass disease, glass illness, crizzling and weeping.

These symptoms reflect the chemical instability of certain types of glass as a consequence of their manufacturing technology, such as over-purification of raw materials or poor batch formulation. A wide range of approaches have been explored to mitigate such problems. These issues have been reviewed in detail (22). It has been found that there are significant discrepancies between those that have been recommended in the past and those in the current conservation literature. This indicates the need for further research.

6.2.8 *Epoxy-Amine Resins Used for Restoration*

The cure kinetics of two epoxy/amine resins, Araldite® 2020 and AY103-HY956, widely used as adhesives in the restoration of works of art from glass or ceramic was investigated using FTIR spectroscopy (23).

These resins are two-part adhesives, consisting of a resin, based on a diglycidyl ether of bisphenol A, cf. Figure 6.1, and a curing agent which is either a cycloaliphatic amine (isophorone diamine) for Araldite 2020, or a mixture of three aliphatic amines in HY956, cf. Figure 6.2.

The study was based on the collection of infrared spectra, in the range of 4000–600 cm^{-1}, of mixtures of resin and hardener at different

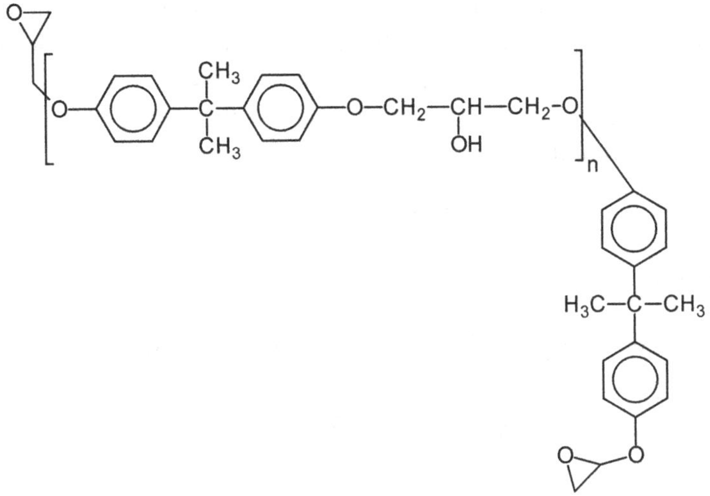

Figure 6.1 Diglycidyl ether of bisphenol A.

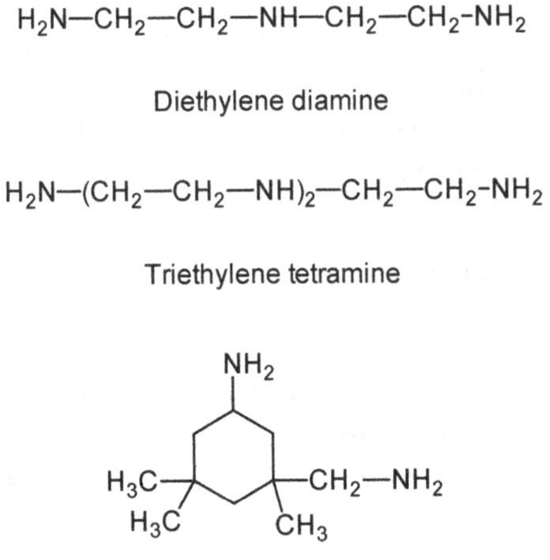

Figure 6.2 Aliphatic amines.

proportions and isothermal temperatures of 22–70°C as a function of curing time.

A kinetic model was used to simulate the experimental data using two kinetic rate constants. Diffusion control was incorporated to describe the behavior of curing at high degrees of conversion.

A general set of equations that describe the epoxy-amine curing involves the reactions of (23):

1. Primary amine addition,
2. Secondary amine addition, and
3. Etherification reaction.

These reactions yield the following kinetic equation (24,25):

$$\frac{d\alpha}{dt} = (k_1 + k_2\alpha)(1 - \alpha)^2 \quad . \tag{6.5}$$

Here, α is the conversion of the epoxy groups and the rate constants k_1 and k_2 reflect the catalysis by groups initially present in the resin and the catalysis by the hydroxyl groups newly formed in the course of the reaction.

From fitting to experimental data the kinetic and diffusional parameters were estimated, together with the activation energies of the kinetic and autocatalytic rate constants. It was found that higher degrees of curing could be obtained at higher temperatures and increased amounts of hardener. Differences in the performance of the two adhesives have been explained based on the type of the amines that were used as curing agents (23).

6.2.9 *Potash Glass Corrosion*

One of the main causes of the deterioration of silicate glasses, in particular in medieval stained glasses of potassium-rich compositions, is the humidity present in their environment.

Model glasses were produced for corrosion tests, with compositions analogous to the ones of the medieval stained glasses from the 15th century from Batalha Monastery. The compositions contained 56 mol% SiO_2, 24 mol% CaO and 20 mol% K_2O.

The corrosion processes of potash glass surfaces in contact with aqueous solutions were studied using ion beam spectroscopic techniques, optical microscopy and FTIR spectroscopy (26).

In order to elucidate the progress of corrosion, the surface studies were combined with the evaluation of the changes that were taking place in the aqueous solution. Thus, the hydrogen profiles obtained from the attacked surface region of glass were compared with the variation of pH values of the aqueous solutions, in order to get information about the ion exchange processes.

The change of the pH measured when glass samples were under immersion were found to increase with time, as shown in Figure 6.3.

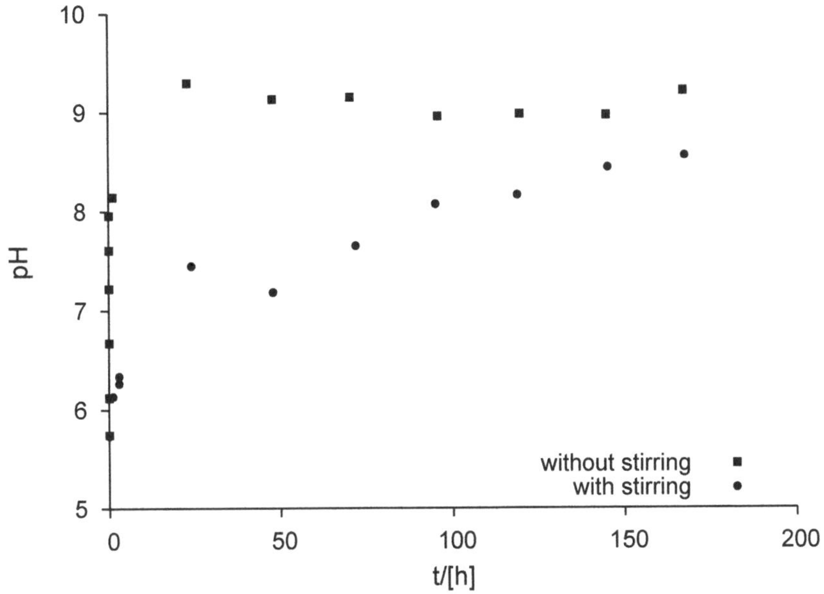

Figure 6.3 Change of the pH with time, without and with stirring (26).

It has been found that pH values may also be a good parameter for studying the kinetics corrosion under high humidity conditions. Tests with and without stirring of the aqueous solution showed different surface morphologies that could be associated with different corrosion rates. Experiments with stirring resulted in a more fine grained and uniform morphology of the corrosion products and a higher corrosion rate (26).

The observed pH evolution can be interpreted as follows: In the first step a selective leaching of K^+ and Ca^{2+} ions occurs. During this process a layer with a more open structure is formed, and therefore the movement of the ions in the network becomes faster. In the second step the formation of molecular water inside the leached layer occurs.

This step corresponds to a higher rate of pH variation because the leaching reactions occur at a faster rate in the open structure that was previously formed. Finally, in the last step of this process, an equilibrium between the leaching of the alkali and alkaline-earth ions takes place and silicic acid is formed that enters into the solution. In the course of this process the pH varies only slightly (26).

6.2.10 *Zinc Treatment on Float Glass*

Float glass is susceptible to corrosion and staining by water (27). It has been known for more than 60 years that treating the glass with zinc may successfully inhibit this process.

The effect of zinc treatment on the corrosion flux of water into float glass has been measured by a gravimetric technique and was compared with untreated samples. The corrosion flux after 300 *min* for an untreated sample was 11.8 times higher than that of a zinc-treated sample exposed at 15°C, 9.25 times higher at 20°C and 2.85 times higher at 30°C. Also, surface analysis techniques were used to identify the presence of zinc on the treated samples (27).

6.2.11 *Sol-Gel Silica Coating*

The applicability and the protective effect of a sol-gel silica coating on potash-lime-silica glasses of low durability has been investigated (28). Glasses of this type were applied during the Romanesque and Gothic periods for manufacturing stained glass objects still present in historic buildings such as cathedrals or smaller churches north of the Alps (29). The high potash content and the low silica amount of these glass types result in a low chemical stability and lead to degradation processes on the glass surfaces.

Two synthetic glasses with compositions close to the historical stained glass were used for the studies (28).

The sol-gel process is a method for producing solid materials from small molecules (30). A silica alcoholic suspension based on sol-gel technology (SIOX-5) was used to obtain a highly pure silica protective film (31).

The sol-gel coating does not alter the optical characteristics of the surface and is compatible with the substrate since its composition and chemical-physical properties are similar to the glass it is applied on (28). In addition, the coating can be easily applied and does not require any heating treatments after the application.

To check the effect of the sol-gel coating, leaching tests with different times and pH values were performed both on the coated and the uncoated parts. The protective effect was verified using light microscopy. The ion migration (e.g., K, Na, Ca and Mg) was assessed using ToF-SIMS analysis.

In summary, the coating also protects the glass from the formation of cracks after severe treatments that would have caused serious damages on the uncoated glass (28).

Also, hybrid organic-inorganic silica protective coatings on glass have been reported (32). Advanced hybrid organic-inorganic coatings with strong weatherability have been presented. Various sols have been prepared by mixing a silica-based inorganic matrix, tetraethyl orthosilicate, with various quantities of silica alkoxides that were functionalized with certain organic groups. Tetraethyl orthosilicate is shown in Figure 6.4.

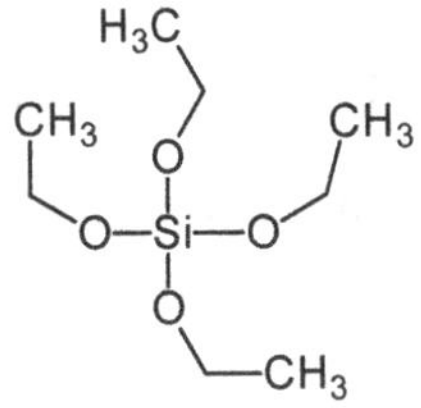

Figure 6.4 Tetraethyl orthosilicate.

In comparison to commercial hybrid silica products, alkyl-functionalized and methacryloxy-functionalized silica coatings particularly displayed improved homogeneity, elasticity and barrier properties (32).

6.2.12 *Hybrid Sol-Gel-Based Coatings*

Medieval glass is commonly attacked by atmospheric pollutants conveyed by water, triggering the corrosion process (33, 34). The current conservation strategies aim to maintain the glass in its original context. Therefore, it is necessary to protect it from further degradation. The sol-gel technology is very effective for the preparation of protective films, using silicious alkoxide precursors that are chemically similar to the substrate.

Water repellent hybrid sol-gel coatings have been described that are made from tetraethyl *o*-silicate with different quantities of Si-alkoxides, functionalized with various alkyl groups. These were octyl triethoxysilane, hexadecyl trimethoxysilane, 3-(trimethoxy silyl)-propyl methacrylate), trimethyl ethoxysilane, and methyl triethoxysilane.

The coatings were deposited using the dip coating technique. The coated materials were characterized by UV-VIS and FTIR spectroscopy.

Measurements of the contact angles showed an overall homogeneity of the coatings and indicated an improved water repellency when the silicones were functionalized with long alkyl chains.

Electron microscopy images showed comparatively even, crack-free coatings, in particular with formulations of higher organic contents. A significant decrease in the contact angle was observed for samples that were treated with very high SO_2 concentrations (33).

6.2.13 *Cyclododecane as Whitening Spray*

The problem of acquiring archaeological artifacts using triangulation-based 3D laser scanners for reflective or refractive surfaces has been handled (35).

The artifacts are mostly made of glass or polished metal. The unique and fragile nature of archaeological artifacts adds an extra constraint to the acquisition process. The use of industrial whitening sprays should be avoided, due to the physicochemical processes required to clean the surface after scanning and because the chemical properties of these sprays may damage the original object. As an alternative to conventional whitening spray the use of cyclododecane has been suggested as a whitening spray. Cyclododecane is shown in Figure 6.5.

Due to its chemical stability and the fact that it sublimes at room temperature, together with its good film-forming capabilities, evaluation tests have been presented to prove that the error introduced by

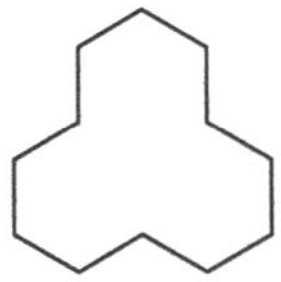

Figure 6.5 Cyclododecane.

the opaque thin layer created on the surface of the artifact is smaller than the accuracy of the 3D scanner and, thus, no acquisition errors are introduced (35).

6.3 Production Practices

6.3.1 Production Practices in Medieval Stained Glass Workshops

The production practices in medieval stained glass workshops have been described in detail (36). Two examples of technical and physical evidence of early medieval stained glass workshop practices have been examined.

A group of 13 panels of French grisaille from various buildings, which date from the 1250s to the 1270s, and panels representing three standing bishops from the band windows in the Norman abbey church of Evron, which date from the second decade of the 14th century, have been assessed. All of the panels are now housed in the Glencairn Museum in Bryn Athyn, Pennsylvania.

The first group seems to an example of ghost images-corrosive imprints resulting from vaporized deposits transferred from one piece of glass to another when they were fired in the kiln, suggesting that the panels were stacked in the kiln.

The second group refers to a number of painted marks that would facilitate sorting and assembling pieces of repetitively designed architecture for leading an individual panel. Such ghost images and sorting marks appear to document the desire to facilitate the assembling and arranging of repeated, modular pieces of fired ornamental glass in order to save time and money in the production of stained glass windows (36).

6.3.2 *Coloring Methods of Old Glass*

Glass fragments from the seventh and eighth century AD were excavated in the Crypta Balbi in Rome (37). These fragments were studied in order to detect the various agents used for color development and opacification (37).

Reflectance spectra on the powdered samples showed the relevance of Fe(II), Fe(III), Mn(III), Cu(II), and Co(II) ions for the color hues. The effect of the atomic ratio of Mn Fe for glass color has been assessed.

It has been concluded that the medieval glassmakers in Italy could make a wide range of colors by exploiting the presence of iron and manganese as contaminants of sand and flux and also by controlling the amount of oxygen guided into the furnace.

The presence of calcium antimonate was detected in white, blue, and blue-green fragments. Elemental copper was detected in a sample from red glass (37).

6.3.3 *Reverse Painting on Glass*

Reverse painting on glass is a technique which consists of applying a cold paint layer on the reverse-side of glass (38). The main challenge facing these artworks is the fragile adhesion of the pictorial layer. A simple movement can modify the appearance of the painting. The pigments used in the study are summarized in Table 6.3.

Table 6.3 Pigments (38).

Pigment name	Chemical name	Thickness/[μm]
Prussian blue	Iron(III) hexacyanoferrate(II)	32
Malachite	Basic copper(II) carbonate	27
Green earth	Basic copper(II) carbonate	47
Indigo	Natural vegetable pigment	40
Vermilion	Mercury(II) sulfide	17
Dragon's blood	Plant resin: cinnabarone	104
Red ochre	Iron(III) oxide (clay, silica)	34
Yellow ochre	Goethite, clay, silica	79
Lead white	Lead(II) carbonate	27

The adhesion parameters of pigments on glass and the comparison between different pigments have been investigated. The relationships between the binder of linseed oil with pigments and the glass with or without the use of an adhesive are studied (38).

The use of a sessile drop device, optical microscopy, SEM, a surface 3D profiler and a pencil hardness scratch tester have been found to be necessary to obtain comparative data of the adhesion of the pictorial layer.

The effect of two adhesives were compared, namely ox gall and gum arabic. It has been shown that the adhesion is not only dependent on the physical parameters but also that possible chemical reactions can influence the effectiveness.

Also, a treatment based on extreme-humidity storage has shown the weakness of some pictorial layers. The humidity treatments were found to be very destructive for the glass slides only covered with adhesives (38).

6.4 Special Uses of Glass Materials

6.4.1 *Medieval Glass Windows*

Medieval stained glass windows are constructions of extreme fragility (39). Each consists of numerous pieces of colored glass of varying chemical composition held together in a firm but elastic network of cames, or thin lead strips with grooves to hold the glass.

The cames follow the lines of the composition, producing a unified, mosaic-like image.

In most cases the glass has been colored with metal oxides, which were added to the molten mass in the course of manufacturing.

An exception is flashed glass, in which a thin film of usually red color was fused onto the clear base glass material. In this case, detailed effects can be achieved by grinding away certain parts of the colored film.

Medieval glass was made from local raw materials, i.e., a mixture of sand and ash from beech wood or fern. Such a mixture shows the advantage of easy melting. However, the glass is soft. This property makes it susceptible to weathering (39).

6.4.2 *Church Windows*

The extant stained glass of Canterbury Cathedral dates before 1600, and includes glass in its collections. The history and documentation of the building has been discussed as well as the dating of the windows, their iconography, style, and techniques used (1).

The glass of the late 12th century building comprises the famous typological series, completely reconstructed with a new edition and translation of the inscriptions recorded.

Also, the extended genealogy of Christ from the clerestory and the miracles of Thomas Becket in the Trinity Chapel have been described. Among the later glass, the so-called Royal Window of the Matyrdom transept is redated by heraldry to the mid-1480s. It is viewed as a memorial to Edward IV and his line, rather than his gift. The west window, perhaps a gift of Richard II, has an extraordinary quality.

6.4.3 *Archaeological Glass*

Glass artifacts that are successfully recovered and preserved have the potential to contribute to the archaeological knowledge in many ways (4). However, the conservation and preservation of archaeological glass may emerge as a challenging task. Some glasses are inherently less stable than others, and some burial environments are more deleterious. In addition, surface weathering of archaeological glass samples is a common effect.

A successful preservation plan for glass must contain both active and preventive conservation and actually starts already before its excavation.

Naturally, because glass is a fragile material, careful handling, both at the time of excavation and subsequently, is important for its preservation. Once a glass artifact has been excavated, the level of deterioration, physical damage, and the stability of the glass itself must be evaluated, and steps to improve the condition of the artifact should be taken if necessary (4).

References

1. M.H. Caviness, *The Windows of Christ Church Cathedral, Canterbury*, Oxford University Press, London, 1981.
2. R. Marks, *Stained Glass in England During the Middle Ages*, Routledge, London, 1993.
3. S. Davison, *Conservation and Restoration of Glass*, Butterworth-Heinemann, Oxford, 2nd edition, 2003.
4. S. Davis and C. Chemello, *Glass: Conservation and preservation* in C. Smith, ed., *Encyclopedia of Global Archaeology*, pp. 3047–3050. Springer, New York, NY, 2014.
5. M. Tutton, *Windows: History, Repair and Conservation*, Taylor and Francis, Hoboken, 2015.
6. N. Zacharias and E. Palamara, *Glass corrosion: Issues and approaches for archaeological science* in F. Gan, ed., *Recent Advances in the Scientific Research on Ancient Glass and Glaze*, chapter 12, pp. 233–248. World Century Publishing Corp, Hackensack, NJ, 2016.
7. M.A. Castro, F.J. Pereira, A.J. Aller, and D. Littlejohn, *Atmospheric Environment*, Vol. 98, p. 41, December 2014.
8. H.R. Golob and E.L. Swarts, *Journal of the American Ceramic Society*, Vol. 67, p. 564, 1984.
9. A. Ceglia, W. Meulebroeck, K. Baert, H. Wouters, K. Nys, H. Thienpont, and H. Terryn, *Surface and Interface Analysis*, Vol. 44, p. 219, 2012.
10. W. Meulebroeck, P. Cosyns, K. Baert, H. Wouters, S. Cagno, K. Janssens, H. Terryn, K. Nys, and H. Thienpont, *Journal of Archaeological Science*, Vol. 38, p. 2387, 2011.
11. D. Lauwers, A. Candeias, A. Coccato, J. Mirao, L. Moens, and P. Vandenabeele, *Spectrochimica Acta Part A: Molecular and Biomolecular Spectroscopy*, Vol. 157, p. 146, 2016.
12. J.T. van Elteren, A. Izmer, M. Šala, E.F. Orsega, V.S. Šelih, S. Panighello, and F. Vanhaecke, *Journal of Analytical Atomic Spectrometry*, Vol. 28, p. 994, 2013.
13. E. Frodl-Kraft, *Studies in Conservation*, Vol. 20, p. 105, January 1975.
14. T. Palomar, *International Journal of Applied Glass Science*, 2016.
15. S. Davison, *Studies in Conservation*, Vol. 29, p. 191, 1984.
16. W.E. Krumbein, C.E. Urzì, and C. Gehrmann, *Geomicrobiology Journal*, Vol. 9, p. 139, 1991.
17. A. Rodrigues, S. Gutierrez-Patricio, A.Z. Miller, C. Saiz-Jimenez, R. Wiley, D. Nunes, M. Vilarigues, and M.F. Macedo, *International Biodeterioration & Biodegradation*, Vol. 90, p. 152, May 2014.
18. Wikipedia, Lichen — wikipedia, the free encyclopedia, 2017.

19. M.R.D. Seaward, *Lichens as agents of biodeterioration* in D.K. Upreti, P.K. Divakar, V. Shukla, and R. Bajpai, eds., *Recent Advances in Lichenology: Modern Methods and Approaches in Biomonitoring and Bioprospection, Volume 1*, pp. 189–211. Springer India, New Delhi, 2015.
20. L.D. Ferri, P.P. Lottici, and G. Vezzalini, *Corrosion Science*, Vol. 80, p. 434, 2014.
21. R. Abd-Allah, *Journal of Cultural Heritage*, Vol. 14, p. 97, 2013.
22. J.J. Kunicki-Goldfinger, *Studies in Conservation*, Vol. 53, p. 47, 2008.
23. E.G. Karayannidou, D.S. Achilias, and I.D. Sideridou, *European Polymer Journal*, Vol. 42, p. 3311 , 2006.
24. K. Horie, H. Hiura, M. Sawada, I. Mita, and H. Kambe, *Journal of Polymer Science Part A-1: Polymer Chemistry*, Vol. 8, p. 1357, 1970.
25. C.W. Wise, W.D. Cook, and A.A. Goodwin, *Polymer*, Vol. 38, p. 3251, 1997.
26. M. Vilarigues and R.C. da Silva, *Journal of Non-Crystalline Solids*, Vol. 352, p. 5368, December 2006.
27. J.N. de Carvalho, J.A.S. Cleaver, N.F. Kirkby, and P. Holmes, *Glass Technology-European Journal of Glass Science and Technology Part A*, Vol. 55, p. 14, 2014.
28. M. De Bardi, H. Hutter, M. Schreiner, and R. Bertoncello, *Journal of Non-Crystalline Solids*, Vol. 390, p. 45, April 2014.
29. M. De Bardi, H. Hutter, M. Schreiner, and R. Bertoncello, *Heritage Science*, Vol. 3, July 2015.
30. Wikipedia, Sol-gel — wikipedia, the free encyclopedia, 2017. [Online; accessed 3-January-2017].
31. C.J. Brinker and G.W. Scherer, *Sol-gel Science: The Physics and Chemistry of Sol-gel Processing*, Academic Press, London, New York, 2013.
32. B. Holubová, Z.Z. Cílová, I. Kučerová, and M. Zlámal, *Progress in Organic Coatings*, Vol. 88, p. 172, November 2015.
33. L. de Ferri, P.P. Lottici, A. Lorenzi, A. Montenero, and G. Vezzalini, *Journal of Sol-Gel Science and Technology*, Vol. 66, p. 253, 2013.
34. L. de Ferri, P.P. Lottici, A. Lorenzi, A. Montenero, and G. Vezzalini, Hybrid sol-gel based protective coatings for historical window glasses, in M.A. Rogerio-Candelera, M. Lazzari, and E. Cano, eds., *Science and Technology for the Conservation of Cultural Heritage*, pp. 231–234, Leiden, The Netherlands, 2013. Proceedings of the Second International Congress on Science and Technology for the Conservation of Cultural Heritage, CRC Press.
35. C. Diaz-Marin, E. Aura-Castro, C. Sánchez-Belenguer, and E. Vendrell-Vidal, *Journal of Cultural Heritage*, Vol. 17, p. 131, 2016.
36. M.W. Cothren, *Journal of Glass Studies*, Vol. 41, p. 117, 1999.
37. P. Mirti, P. Davit, and M. Gulmini, *Analytical and Bioanalytical Chemistry*, Vol. 372, p. 221, 2002.

38. M. Bayle, D. Waugh, B. Colston, and J. Lawrence, *Applied Surface Science*, Vol. 357, Part A, p. 293, 2015.
39. G. Frenzel, *Scientific American*, Vol. 252, p. 6, 1985.

7

Archaeological Metals

7.0.4 Analytical Methods

7.0.5 Dating Archaeological Lead Artifacts

Dating is important in archeometry. However, in the course of dating significant problems emerge, in particular in the case of metallic samples.

The ^{14}C method is sometimes used with the traces of charcoal found in iron objects. However, the main techniques for absolute dating in archeometry, i.e., the radiocarbon method and the uranium-lead radioactive series method, are not used for dating metals (1).

The Meissner fraction in the superconducting state of lead archaeological artifacts can be used for accessing the mass of the uncorroded metal in the sample. From the total mass of the sample, the mass of all the corrosion products can be established. It has been found that this mass correlates with the archaeological age of the lead artifacts over a time span of around 2500 years.

This chemical dating method is applicable to lead artifacts buried in soils with a pH greater than 6.5. In such soils the corrosion process proceeds very slowly and the corrosion products, mainly PbO and $PbCO_3$, accumulate over hundreds of years (1).

Another methodology for dating archaeological lead artifacts has been presented (2, 3). The method is based on the voltammetry of microparticles. The basic issues this method has with respect to archeometry have been critically reviewed (4). Originally this procedure was developed by Scholz in the 1980s (5).

This methodology is based on the comparison of the height of specific voltammetric features from PbO_2 and PbO corrosion products formed under long-term alteration conditions.

In the course of voltammetry, basically two electrochemical processes are observed: Process 1 is attributed to the reduction of the primary, nonporous PbO patina. Process 2 is attributed to the reduction of the porous layer resulting from kinetic passivation.

In the course of a voltammetric scan, the peak areas of the electric current A_1 and A_2 are measured belonging to the two above-mentioned processes.

Based on the assumption that the corrosion process arises from the local existence of an electrochemical cell which consists of a junction of metallic oxide to a metal surface and the oxygen/water couple, the corrosion rate follows a potential law (1). Thus the peak areas should satisfy the relationship:

$$\frac{A_2}{A_1} = G_1 + [G_2(1+\alpha)t]^{(\frac{1}{1+\alpha})} \tag{7.1}$$

In Eq. 7.1, G_1 and G_2 are constants depending on the mass of the primary patina by surface unit. This can be taken as essentially identical for all the samples under investigation. Therefore, Eq. 7.1 should predict a linear relationship between the ratio $\frac{A_2}{A_1}$ and $t^{\frac{1}{1+\alpha}}$, for an appropriate selected value α.

So, the relative amount of the oxides can be estimated from polarization curves and electrochemical impedance spectroscopy (EIS). This approach is complemented by additional data using square wave voltammetry data.

The calibration of the method was performed using a series of well-documented lead pieces from the funds of different museums of the Comunitat Valenciana (Spain) covering the fifth century BC to the present time.

A plot of the ratio of the areas against time using $\alpha = 0.07$ is shown in Figure 7.1.

The methodology has also been used for the authentication of archaeological lead (6). An Iberian lead plate from the Tossal de Sant Miquel (Llíria, Spain) site was investigated.

Nanosamples from the archaeological artefact were mechanically transferred to a graphite pencil electrode in contact with aqueous

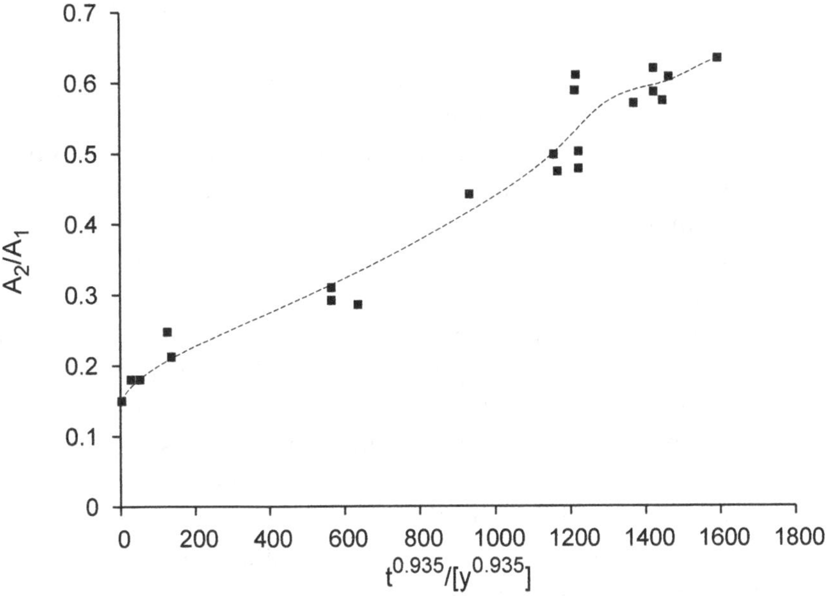

Figure 7.1 Plot of the ratio of the areas against time using $\alpha = 0.07$ (3).

buffers. Three diagnostic criteria for authentication have been described based on the appearance of (6):

1. Oxidative dissolution signals for trace metals like copper, arsenic, antimony and, often, tin and silver accompanying stripping peaks for lead,
2. Peak potential shifts for reduction peaks for patination products, and
3. The presence of reduction peaks for PbO_2.

7.0.6 Lead Isotopic Measurements

The isotopic analysis of lead on metal artifacts may help to elucidate their provenance. Useful and reliable data require analytical protocols that ensure accuracy and monitor precision.

One problem may be that the interpretation will be ambiguous because natural ore deposits frequently have overlapping lead isotopic compositions.

However, mostly, the provenances can be narrowed down. So, for the Alpine region, a pattern of small, strongly localized mine exploitations supplemented by some long-distance trading, can be found.

The methods of lead isotopic measurements and their interpretation, as well as certain examples, have been described in detail (7).

For example, the four isotopes of Pb can be combined into 24 different isotope ratios, at most three of which are truly independent. As an example, the triplet (207Pb/206Pb, $^{207}Pb/^{204}Pb$, $^{204}Pb/^{208}Pb$) contains three independent ratios, while ($^{207}Pb/^{206}Pb$, $^{207}Pb/^{204}Pb$, $^{204}Pb/^{206}Pb$) only contains two.

Each sample can be plotted in a three-dimensional graph whose axes are the three chosen independent ratios. Among all the possible triplets of independent isotopic ratios, the best candidates for the three axes are triplets of common-denominator ratios because only in that case do mixtures between samples lie on a straight tie line between the points of the samples (8).

Lead isotopes combined with trace element data represent a powerful tool for non-ferrous metal provenance studies. Nevertheless, unconsidered geological factors and archaeological data, as well as ignored analytical procedures, may substantially modify the interpretation of the isotopic and trace element signature obtained as a potential ore candidate (9).

7.0.7 *Archaeometallurgical Analysis*

7.0.7.1 *Byzantine Shipwreck*

The Byzantine period Dor 2006 shipwreck was discovered in the Dor Lagoon in Israel (10). The shipwreck was dated between the mid-6th and the first quarter of the 7th centuries AD. Among the remaining samples of metal objects, two coins, a lead sheet, and concretions of nails and a sickle were described.

Metallurgical nondestructive and destructive analysis of the metal remains demonstrated that the coins were made from a copper lead alloy with a high concentration of lead, produced by casting, and later stamped with coin die.

The lead sheet was produced by casting and later shaped into its final form. Scanning electron microscope (SEM), energy disperse spectroscopy, X-ray diffraction (XRD), and other methods revealed that the nail remains were essentially completely corroded iron.

The analysis of the completely corroded sickle rests revealed that it was made from bloomery iron. This suggested that the use of lead in the coins and tin in the lead sheet was due to economic constraints.

Iron nails, although widespread, also indicate a preference for a cheaper metal than copper alloy. The lead isotope signature of the lead sheet indicates that the source of the ore was Greece or the Taurus Mountains, which may hint at the origin of the ship (10).

7.0.7.2 *Numismatic Artifacts*

The first study on archaeological metal objects was conducted at the beginning of the 19th century by M. H. Klaproth (11), who published the study of a Roman coin, with the first quantitative analysis of an alloy.

Several coins have been analyzed using energy dispersive X-ray spectroscopy (EDX), XRD and other related techniques (12).

7.0.8 Dating of Archaeological Copper Samples

A methodology for dating copper/bronze archaeological objects aged under atmospheric environments using EIS has been described (13). The method is based on the measurement of resistance associated with the growth of corrosion layers in EIS recorded upon immersion of the pieces in mineral water and applying a bias potential for the reduction of dissolved oxygen.

Theoretical expressions for the time variation of such resistance following a potential rate law have been presented. Equivalent expressions were derived and applied for estimating the variation of the tenorite/cuprite ratio from their specific voltammetric signals using voltammetry data of the microparticles. In addition, calibration curves were constructed from a set of well-documented coins.

It is assumed that under ordinary atmospheric corrosion conditions, copper and bronze samples form a primary patina of cuprite, Cu_2O, which subsequently grows and will be partially oxidized to tenorite CuO, thus forming a secondary patina.

This oxidation is a thermodynamically spontaneous process favored by the prolonged contact of the artifacts with a carbon dioxide-rich atmosphere as well as with calcareous materials (14–16). This kinetic model is based on the following assumptions (17):

1. There are simultaneous or consecutive interfacial reactions occurring at the metal primary patina and primary patina and secondary patina boundaries,
2. The kinetic potential laws are valid for such reactions. This hypothesis is supported by the literature on long-term corrosion of metals (18–21).

Basically the kinetic equation of each patina formation looks like:

$$\frac{dy}{dt} = ky^{-a} \quad . \tag{7.2}$$

Here, y is the thickness of each layer. These equations can be combined to get a more complicated equation as in the case of combined followed reactions (13). The kinetic equations of the formation of the patina have been detailed elsewhere (13).

7.0.9 Laser-Induced Breakdown Spectroscopy

Various materials immersed in seawater have been characterized by means of Laser-induced breakdown spectroscopy (22). The plasma emission was produced by a Q-Switched Nd:YAG laser operated at 1064 *nm* in a dual pulse mode. Data acquisition and processing could be optimized for better signal control and also to improve the detection threshold.

Quantitative chemical analysis was performed on submerged bronze samples. Calibration curves have been obtained with standards of similar matrix composition. The optimized experimental conditions for the analysis of different materials are collected in Table 7.1.

7.0.10 Voltammetric Analysis

7.0.10.1 Roman Bronze Coins

Voltammetry of microparticles and EIS techniques, together with SEM-EDX and Raman spectroscopy, were applied to a set of 15 Roman bronze coins and one Tessera from the temple of Magna

Table 7.1 Optimized experimental conditions for the analysis different materials (22).

Material	Trigger Delays/[ns]	Pulse Energies/[mJ]	Laser shots	Acquisition gate/[ns]
Stainless steel			20	
Iron–clean			20	
Bronze	$t_1 = 150$	$E_1 = 120$	20	400
	$t_2 = 55$	$E_2 = 160$		
Corroded iron			100	
Precious alloys			1	
Marble, rock	$t_1 = 145$	$E_1 = 82$	50	600
	$t_2 = 55$	$E_1 = 180$		

Mater (Rome, Italy) (23). This archaeological site dates back between the second half and the end of the 4th century AD.

Characteristic voltammetric patterns for cuprite and tenorite for submicrosamples of the corrosion layers of the coins deposited onto graphite electrodes in contact with 0.10 M $HClO_4$ aqueous solution yielded a grouping of the coins into three main groups.

This grouping could be confirmed and refined using EIS experiments of the coins immersed in air-saturated mineral water by the reduction of dissolved oxygen as a redox probe. The electrochemical grouping of coins corroborated the complex stratigraphy of the archaeological site. Also, it suggested the reuse of the coins during the later periods possibly due to economic issues related to the decline of the Roman Empire (23).

7.0.10.2 Polish Coins

A series of 20 denarii from Boleslaw the Brave (992–1025) and Mieszko II Lambert (1025–1034), corresponding to the beginning of the Polish state, were studied using voltammetry of the immobilized particles (24). The experiments were applied to nanosamples of the corrosion layers of the coins in contact with aqueous acetate buffer.

Well-defined responses were found mainly corresponding to the corrosion products of copper and lead. The voltammetric responses, combined with X-ray fluorescence analysis spectroscopy experiments performed on the same set of coins, together with focusing ion beam-field emission scanning electron microscopy on silver coins from the 19th century, supported the hypothesis that two different metal sources were used in the former historical period. So it was suggested that the coins were produced in three different mints (24).

7.0.11 *Energy Dispersive X-ray Fluorescence Analysis*

Energy dispersive X-ray fluorescence (EDXRF) is widely used in the study of archaeological metal artifacts, heritage and art history, where the fragile nature of the objects requires the use of noninvasive techniques such as EDXRF (25). This technique is fast and very affordable.

7.0.11.1 *Late Bronze Age Metal Hoards from Central Portugal*

An EDXRF analysis of eleven copper-based artifacts from Late Bronze Age metal hoards from Central Portugal has been detailed. The EDXRF measurements were carried out with an X-ray tube with a Mo anode and a commercial Si-PIN detector.

The data acquisition was performed at small distances between the X-ray window, the sample and the detector. Both patinated and polished areas have been analyzed: The relative composition of the artifacts was obtained from the fluorescence spectra.

Cu was used as reference element. The measured intensities of Fe and Sn with respect to those of Cu were corrected by introducing a calibration coefficient.

The results of the analysis of the polished areas indicated that all eleven artifacts were manufactured with a high pure tin-bronze alloy and showed a quite similar composition, independent from their provenance and typology. Actually, no significant differences were observed either with regard to the main elements or to the impurities. The Sn content varies from 6.33% to 12.55%, with an average composition of 10.18% (25).

From a structural point of view, a relative weight of Sn of around 10% provides a bronze alloy with excellent thermomechanical properties (26).

The presence of Fe in the protohistoric copper metallurgy of Western Iberia can provide important information about the nature of the smelting process: Low rates of iron suggest a primitive metalwork, i.e., the use of reaction vessel or crucibles, low temperatures and poor reducing conditions, whereas its increase reveals the adoption of a more sophisticated and efficient smelting process.

Actually, at the beginning of the Iron Age, with the Phoenician colonization in the south of the Iberian Peninsula, the introduction of technological changes in the furnaces produced a significant increase of iron in a copper-based alloy (27).

7.0.11.2 *Ancient Copper and Bronze Artifacts from Albania*

Ancient copper and bronze artifacts from different Albanian archaeological sites have been analyzed by X-ray fluorescence analysis to determine the type of the alloys and the elemental composition. In addition, optical microscopy was used for the investigation of corrosion products and microstructure of the bulk metal.

Three bronze coins were excavated in Dyrrahu, composed of copper, tin and lead. Also, nails, excavated in Dyrrahu, and another one found inside an amphora in an ancient ship close to the coast of Saranda, have been found to be composed of copper. Nondestructive elemental analysis of the artifacts was performed using a portable X-ray fluorescence system.

The strata-graphical morphology and microstructure images from microscopic examinations allowed the identification of different corrosion products and the diverse stages of metallurgical and mechanical treatment during their manufacture. The results of the characterization can be summarized as (28):

- Both nails are composed of pure copper.
- The strata-graphical morphology of the ancient copper in the nails has a sandwich structure. In the case of the nail from Dyrrahu, the primary patina contains layers of tenorite and cuprite, growing on the surface or into the metal grains, while the secondary patina contains layers of malachite. In the case of the nail from Saranda, there exists only the primary patina which contains layers of tenorite and cuprite.
- In the microstructure of the two nails, twin lines can be clearly seen, which traverse part or all of the individual grains of metal. Some of the twin lines are perfectly straight,

few twins are curved. The recorded Vikers microhardness values on the bulk of copper nail from Dyrrahu are in the range of 44.4 *HV* up to 57.1 *HV*. This confirms that the nails are roughly cast to shape, and are worked to finish at hot temperature, where they are annealed.

- The alloy of coin from Dyrrahu is a leaded bronze with 78.2% Cu, 10.1% Sn, 11.2% Pb, and 0.2% As.
- The presence of tenorite and cuprite on the surface and inside the corroded coin is a proof of the authenticity of the coin from Dyrrahu.
- The microstructure of the coin from Dyrrahu shows a recrystallized structure. Many of the twin lines seem quite straight, while others seem curved. In addition, many fine sets of parallel slip lines or strain lines can be seen with extensive evidence of deformation produced in the process of striking of the coin. The curved twins and the sets of slip lines indicate that either hot-working or some cold-working after the final stage of recrystallization may have been used during the striking of coins.
- The recorded Vikers microhardness value on the bulk of leaded bronze of coin alloy are in the range from 90,5 to 117 *HV* with an error of 2 *HV*.

7.0.12 Roughness Estimation of Archaeological Metal Surfaces

By means of a time domain spectrometer, the reflection of terahertz pulses from surfaces have been investigated that exhibit a variable degree of roughness (29).

Traditionally, in the automotive industry, the thickness of multi-layer paintings have to be accurately measured using terahertz spectroscopy in order to assess whether adjacent layers are correctly detached (30).

The study with regard to archaeological metal surfaces was focused on assessing the influence of the surface texture on the amplitude and the shape of the pulses reflected by stratified materials and on exploring the potential of this technique for achieving quantitative information on the roughness of the material interfaces hit by the *THz* beam.

The behavior of the reflected *THz* pulses was assessed by angular measurements on a set of suitable mock-ups. The measurements were carried out on an authentic archaeological Roman coin that exhibited different corrosion situations.

An electromagnetic model was used for estimating the roughness of the outer and the inner surfaces. Comparing the results with those provided by other techniques made it possible to parameterize the surface texture, such as the traditional contact micro-profilometry and the 3D digital microscopy (29).

7.0.13 *Energy Dispersive X-ray Fluorescence Spectrometry*

The uses of EDX as an analytical technique for studies of ancient coins have been reviewed (31). Specific EDX applications in historical studies, studies of the corrosion of coins, and studies of the optimal working conditions of some laser-based treatments for the cleaning of coins have been described.

7.1 Cleaning Methods

7.1.1 *Tarnished Silver*

Tarnishing occurs by the reaction of silver with gaseous reduced sulfur compounds from atmospheric pollution (32). It induces the formation of Ag_2S crystals, which result in a dark appearance.

In order to remove the sulfur tarnishing, various mechanical, chemical and electrochemical methods have been used. In order to assess the effect of the different cleaning treatments on tarnished silver, coupons of pure and sterling silver containing 92% Ag and 8% Cu were subjected to six cycles of an induced tarnishing and cleaning.

The mechanical cleaning methods were soft abrasives and a rubber point mounted on a rotary tool. The mechanical cleaning methods were the use of chelating and acid solutions. As electrochemical cleaning method, potentiostatic reductions were applied (32).

The surface morphology, composition, weight, color and luminosity were evaluated by microscopic and spectrometric methods. It was found that the cleaning effectiveness on tarnished silver depended on the cleaning procedure and the composition of the silver.

Mechanical treatments restored the original visual appearance of silver, but they produced a significant mass loss and a fast re-tarnishing. On the other hand, chemical cleaning methods were fast. However, the surface appeared completely attacked. Finally, electrochemical cleaning did not yield a good result for sterling silver, but it turned out to be an effective cleaning method for pure silver coupons (32).

7.1.2 *Laser Cleaning*

The first report of laser cleaning of metal artefacts was made in the late 1970s by John Asmus (33). He performed cleaning trials on lead samples with calcareous deposits and on corroded bronze (34).

Since then, only a few applications of laser cleaning to metal artworks have been reported. In 1991, experiments with a Q-switched ruby laser on corroded bronze artefacts were described (35). These results demonstrated that long laser pulses are more likely to induce surface melting when irradiating metal objects, So, short laser pulses in the nanosecond range are preferable.

Then, systematic research on the process of laser cleaning of metals was carried out at Loughborough University by using a TEA (transversely excited atmospheric pressure) CO_2 laser.

The effects of laser cleaning have been investigated in samples of various metals such as copper, bronze, steel, lead and brass. These studies demonstrated that laser cleaning is particularly effective for removing paints and stains of organic nature, without damaging the metal substrate (36,37).

A study has been presented where the application of laser cleaning to excavated metal objects was tested and analyzed (34). The effects of different laser sources were investigated and the complete treatment of the objects was classified, operative and aesthetic aspects were also evaluated.

For example, on bronze artefacts, laser cleaning by QS and SFR Nd:YAG lasers produced a selective ablation of the alteration layers that permitted the preservation of the oxidation patina, without exposing the metallic substrate.

Q-switching is a technique by which a laser can be made to produce a pulsed output beam. The technique allows the production of light pulses with an energy in the range of *GW* (38).

The SFR mode stands for short free running with short time pulses. This term means that the laser emission lasts as long as the pumping process is sufficient to sustain laser emission conditions. Typical pulse durations are in the μs to ms range, depending on the pumping source.

Effects of changes in color after laser irradiation could be sometimes observed in the course of cleaning of bronze and copper samples, which occurred on the surface of calcareous accretions. There the green color of the alteration was replaced by a local discoloration, varying from greyish to reddish, depending on the sample. In summary, the results indicated that the laser cleaning process was selective and highly precise, allowing the preservation of fine surface details (34).

7.1.2.1 *Artificially Aged Copper Specimens*

Laser cleaning was carried out on artificially aged copper specimens (39). The results of different laser irradiance values and different number of passes were compared. The cleaned surfaces were initially characterized by optical and scanning electron microscopy, X-ray diffraction and X-ray photoelectron spectroscopy.

The analysis showed that near-IR laser cleaning successfully removed the corrosion products from the surface of the metallic artifacts.

After cleaning at higher irradiance or after repeated cleaning steps, a slight surface alteration was observed. The samples were also characterized by optical fiber spectroscopy. A signature in the spectral reflectance could be used to distinguish between metal, corrosion products layer and cleaned surface (39).

7.1.2.2 *Tarnished Silver Artifacts*

The relevant chemical reactions involved in the tarnishing mechanism of silver are:

$$2Ag + H_2S + \frac{1}{2}O_2 \rightarrow Ag_2S + H_2O \tag{7.3}$$

$$2Ag + OCS \rightarrow Ag_2S + CO \tag{7.4}$$

Ag_2S is a black compound and its formation on the silver surface produces a strong loss of shine and a change of color to a dark appearance. This is unacceptable for historic artifacts.

Laser cleaning of tarnished pure and sterling silver substrates has been tested (40). A nanosecond Q-switched Nd:YAG laser at 1064 *nm*, 532 *nm* and 266 *nm* has been used.

In order to assess the effects associated with cyclic laser cleaning treatments, several cycles of tarnishing followed by laser cleaning were applied on silver coupons that were characterized by gravimetry, colorimetry, scanning electron microscopy, X-ray photoelectron spectroscopy and micro-Raman spectroscopy.

The results showed that none of the three wavelengths are recommended for laser cleaning of pure silver objects, while for sterling silver artifacts, the visible laser wavelength of 532 *nm* seems to be the most appropriate (40).

7.1.3 Plasma Sputtering

On the north coast of present-day Peru, between 50 and 700 AD the Moche civilization flourished. There were sophisticated metalworkers who were considered the finest producers of jewels and artefacts of the region (41).

The Moche metalsmiths developed several alloys of copper with either silver or gold and some silver (42). These alloys were used in a wide range of compositions to produce hammered sheet metal to serve as raw material for the manufacture of silver-looking or gold-looking objects by depletion gilding and electrochemical replacement plating.

In particular, tumbaga is a fabricated alloy of copper and gold that follows a depletion and enrichment coloring process and suffers from corrosion reactions when copper is present in high concentrations. Moreover, they invented a technique of electrochemical deposition of thin silver or gold films onto a copper surface. Gilding was achieved by an electrochemical replacement plating process in which gold and silver are dissolved in an aqueous solution of corrosive materials. The precious metals are then plated from solution onto the copper objects. Most objects were made of sheet metal, shaped using solid metals or wood forms over which sheet metals were carefully hammered; so, duplicate objects of nearly identical size and form were also possible (42).

Plasma sputtering used in an electron-cyclotron-resonance device has been applied to clean archaeological Moche artefacts, unearthed at the Royal Tombs of Sipán (43). The Moche civilization, alternatively, Mochica culture, flourished in northern Peru from 100 to 750 AD (44).

After successful cleaning, the pieces were analyzed by a variety of complementary techniques, i.e., proton induced X-ray emission, Rutherford backscattering spectroscopy, X-ray diffraction, electron microscopy, and inductively coupled plasma mass spectrometry.

With these techniques, it has been possible to determine the profiles of the gold and silver surface layers. Also, elements could be detected that could be relevant in explaining the gilding techniques used by the metalsmiths of the Moche culture (43).

In addition, some 50 metal objects were analyzed with portable equipment which uses EDX (41). This portable equipment is mainly composed of a small-sized X-ray tube and a thermoelectrically cooled, small-sized X-ray detector. Standard samples of gold and silver alloys were used for calibration and quantitative determination.

These artefacts are mainly gold, silver and copper alloys, gilded copper and tumbaga, the last being a poor gold alloy enriched at the surface by depletion gilding, i.e., by removing copper and silver from the surface.

In the case of gold, silver and copper alloys, their composition was determined by the EDX analysis (41,45). In the case of gilded copper or tumbaga, the ratios $Cu(K_{\alpha}/K_{\beta})$ and $Au{-}L\alpha/Cu{-}K\alpha$ could be accurately determined from the X-ray spectra. Thus, clearly gilded copper could be distinguished from tumbaga. Also, the gilding thickness or an *equivalent gilding thickness* could be determined in the case of tumbaga. K_{α} and K_{β} occur when an electron transfers to the innermost K shell from a more energetic orbital (L or M orbital) (46).

The combination of the two ratios is a clear indication of the nature of the alloy, i.e., gold, gilded copper or tumbaga. This allows an accurate measurement of the gilding thickness in the case of gilded copper objects. In the case of tumbaga, the equivalent gold thickness could be measured to be around 2.8 μm. The mean approximate composition of Sipán alloys from three tombs is summarized in Table 7.2.

Table 7.2 Composition of gold objects from different tombs (41).

Tomb	Gold/[%]	Copper/[%]	Silver/[%]
Del Señor de Sipán	68.6	8	23
Del Sacerdote	69	8	23
Other tombs	75	10.5	14.5

7.1.4 Thermochemical Treatment for Iron and Copper Alloys

The physical and chemical processes of corrosion need the conservation of metal archaeological findings (47). Over the past several decades, conservation and restoration techniques of metals have been developed. Actually, these techniques have disadvantages and have posed the issue of an unprogressive coating removal.

A thermochemical treatment of iron and copper base alloys from polymer protective coating has been developed. Comparative experiments have been carried out with several chemicals at 185°C to 200°C. A technique for treating archaeological iron from protective coating containing a butvar-phenolic adhesive has been developed.

Regarding the issues of the process of cleaning objects with adhesive layers, the results are as follows (47):

1. Isopropyl alcohol proved itself as a poor solvent of the polymerized butvar-phenolic adhesive. In order to destroy the chemical bonds of butvar-phenolic adhesive it is necessary to soak objects in alcohol for several days, which does not guarantee the uniform vitrification and embrittlement of the layer of butvar-phenolic adhesive. For this reason, a treatment of the same object has to be repeated for at least 2–3 times. Boiling in alcohol also does not lead to the uniform dissolution of butvar-phenolic adhesive.
2. After soaking the object in acetone for 3 *d* and thermal treating at temperatures of 185°C, 200°C, and 210°C, the polymerized layer of butvar-phenolic adhesive becomes brittle and

the object was cleaned off with metal brushes on the polishing machine, but the dissolution of the layer was observed to be uneven. Boiling in acetone showed a satisfactory result. The layer of the butvar-phenolic adhesive is dissolved evenly over the entire surface of the object after 2–3 *h*.

3. Ethyl acetate, cf., Figure 7.2, immediately reacts with the polymerized layer and allows it to be cleaned off. However, the reaction occurs only with the surface layers and chemomechanical treatment has to be repeated several times. Soaking in ethyl acetate showed a good result; the layer of the butvar-phenolic adhesive dissolves within 5–10 *h* depending on its thickness. However, ethyl acetate is an explosive substance and a poison. Also, the vapor of ethyl acetate irritates the mucous membranes of the eyes and respiratory tract, and the effect of the solvent on the skin causes dermatitis and eczema. The maximum permissible concentration in the air of the working area should be 200 $mg\,m^{-3}$, the flashpoint is 2°C, and the ignition temperature is 400°C.

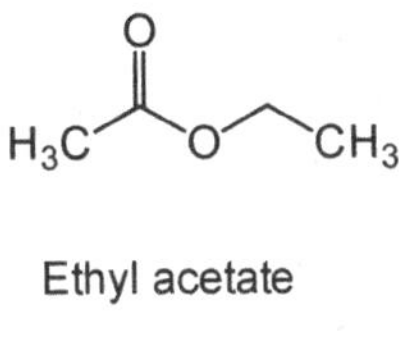

Figure 7.2 Ethyl acetate.

7.2 Special Uses of Metals

7.2.1 *Archaeological Material from Underwater Sites*

Methods and compositions for conserving archaeological materials from underwater sites have been described in detail. The review focusses, among others, on (48,49):

1. Adhesives and consolidants,
2. Conservation of bone, ivory, teeth, and antler,
3. Conservation of pottery,

4. Conservation of glass,
5. Wood conservation,
6. Leather conservation,
7. Textile conservation,
8. Iron conservation,
9. Copper, bronze, and brass conservation,
10. Silver conservation,
11. Lead, tin, and lead alloy conservation, and
12. Gold and gold alloy conservation.

7.2.2 *Bronze Shields*

Atypical formations in the structure of the corrosion crust and in the partially mineralized metallic core have been described (50).

The sample was characterized by optical microscopy and SEM/EDX methods.

The formations resulted during the underground stay of a bronze shield, dated between the 1st century BC and the 1st century AD.

A representative fragment had been chosen from the rim of the shield, which was analyzed by optical microscopy and by electron microscopy coupled with energy dispersive X-ray spectroscopy (50).

In this way, its morphology has been studied, as well as composition and the location of chemical compounds on the surface and inside the bulk formed during the underground stay. This occurred by the processes of chemical and physical alteration, assisted by contamination with structural elements from the site. The processes, i.e., monolithization and mineralization, formed a series of structures consisting of congruent elements and phases with a complex composition.

Such formations are assessed as surface effects generated by exogenous factors and endogenous factors inside the bulk. They are frequently found in ancient bronze objects, such as the exterior flat mole formations and the Liesegang effect (51) in the stratigraphic structure of the bulk.

Some of those structures have atypical characteristics concerning their structure, composition, and formation mechanism, which may be used for archaeological dating (50).

Actually, the experiments performed on the artifact led to the following conclusions (50):

- During the underground stay, the bronze rim that fixed the wood panels of the leather shield underwent processes of alteration that affected its chemical nature, i.e., corrosion, and its physical structure, i.e., erosion, flattening, and cracking,
- In the corrosion crust, basic alloy components (Cu, Sn, and Fe) were found. The components of the three patinas underwent processes of monolithization and crystalline reformation,
- The content of C, S, and P was attributed to the presence of oak wood in the shield and to the leather cover,
- The corrosion crust contained a series of surface effects, such as the Liesegang effect. The flat moles contained circular successive layers that decreased in size from the outside layer, fixed in the corrosion crust at the base,
- The Liesegang effect was found on a limited area, where in the composition of the shield, tin had become concentrated by segregation, which was responsible for forming thin layers of hydrogels.
- The composition and distribution of the structural elements in the object confirm that they come from a shield made of wood covered with leather, which was fixed on the edges with a bronze frame and fastened with rivets.

7.2.3 *Copper and Bronze Axes*

Micro X-ray fluorescence analysis and neutron imaging techniques, such as 2D radiography and 3D tomography, have been used to study four metal axes from the Early/Middle Bronze Age in Western Iberia (52). This period is characterized by a metallurgical change in the use of copper to bronze.

X-ray fluorescence analysis indicated that one of the axes was produced in copper with some arsenic, while the other three axes were produced in a copper-tin alloy, i.e., bronze with variable tin contents and some arsenic and lead.

Neutron radiography and tomography were used to study internal heterogeneities of the axes in a noninvasive way. Namely, the specificities of neutron interaction with matter allow a suitable penetration of comparatively thick copper-based objects in comparison to the use of a conventional X-ray radiography.

Neutron imaging allows the visualization of internal fissures and pores and the evaluation of their distribution, size and shape. Relevant information for the reconstruction of ancient manufacturing techniques could be collected. It was observed that one ax was produced with the mold in an angle of 25 degrees, probably to facilitate the escape of gas during metal pouring. Also, information regarding physical weaknesses of the axes could be collected, thus providing relevant data for the conservation of these artefacts (52).

In summary, it has been concluded that micro-XRF analysis is well suited for the investigation of small surface areas of artifacts that are free of corrosion layers. The results have shown that while one ax was made of copper with arsenic, following the metallurgical tradition from Copper Age until Early/Middle Bronze Age, others were made of bronze, the new alloy that began to be introduced by the Middle Bronze Age in Western Iberia (52).

7.2.4 Coins

Archaeological coins are important as they may deliver to archaeologists valuable information.

7.2.4.1 Authentication of Old Bronze Coins

Comparative results concerning the chemical composition and mineralogical distribution of the main compounds from the structure of the archaeological patina of ancient bronze coins have been presented (53).

These were determined by the corroboration of microchemical methods, colorimetric methods, as well as spectroscopic methods.

The stratigraphic distribution of the chemical components in the structure of the patina is caused by the pedological processes from the archaeological sites. This is representative of the main elements used in the authentication of ancient bronze artifacts.

Three types of products can be differentiated in the structure of the archaeological patina: The first type resulted from corrosion processes, another type from acid-base processes, mainly through hydrolysis and ionic exchange. The third type resulted from contamination processes, such as diffusion, segregation, deposition, and others.

These products are formed in different regions that are characteristic to certain types of patina: The oxides and sulfides, formed in the course of the utilization period, constitute the primary patina. Halogenides, oxyhydroxides, carbonates, sulfates, phosphates, etc., resulted in the last period of utilization and in the first step of the pedological period. These define the secondary patina.

Finally, the products that resulted from physical processes of diffusion, segregation, deposition, and recrystallization form the contamination patina. Their study allows the explanation of the inner mechanisms of formation but also the evaluation of some chronological parameters.

These offer a series of elements that can be used in authentication, such as, the nature of the alloy, the type of the mineral, the provenience of the coin, i.e., period and place of creation (53).

The methods of identification of six ancient Byzantine bronze coins found in the same archaeological site of Nufárul (Tulcea County, Romania), by corrosion product and alloy analysis have been presented (54).

The microstratigraphies, i.e., the layers, were tested with reflection colorimetry, infrared spectroscopy, X-ray diffraction and scanning electron microscopy assisted by X-ray spectrometry.

These tests have rendered evident the stratified morphology of three types of patina. The stratigraphical distribution of the chemical components in the structure of the patina is caused by the pedological (soil) processes at the archaeological sites and can be the main factor used in the authentication of ancient bronze artefacts (54).

7.2.4.2 *Correlation between Corrosion and Soil Composition*

Only a few reports concerning the identification of the most significant parameters that may influence the corrosion of metallic objects are available (55).

A series of ancient coins from the archaeological excavation of Palazzo Valentini (Rome) were collected together with soils, both near and far from them, and analyzed with different analytical techniques, e.g., XRD, electrical conductivity. This was done to establish a correlation between the corrosion products covering the coins and the characteristics of the soil. Also, the content of soluble salts in the water-bearing stratum and surfacing in the archaeological site was measured.

The thus obtained results strongly suggest an influence of the alkaline soils on the formation of a patina. Cerussite, probably due to the circulation of water in layers rich in marble and plaster fragments, was found as the the main corrosion product identified.

By measurements of the electrical conductivity, pH and the content of soluble salts of the washing solutions from both coins and soils, the origin of coins from different stratigraphic units of the site could be identified. The resulting data were tested using cluster analysis and multivariate analysis. In this way, a correlation between part of the coins and the nearby soil samples could be established (55).

7.2.4.3 *Conservation Processes*

Alkaline Rochelle Salt Cleaning. Conservation processes for cleaning marine archaeological silver coins have been described (56). The coins are Shoiba Hoard Coins and originate from the Red Sea near Jeddah, Saudi Arabia.

Five methods were tested for conservation (56):

1. Alkaline Rochelle salt,
2. Alkaline dithionite reduction technique,
3. Electrolytic reduction technique,
4. Alkaline Rochelle salt combined with alkaline dithionite reduction technique, and
5. Alkaline Rochelle salt combined with electrolytic reduction technique.

All the tested cleaning methods were carried out on the coins without mechanical cleaning.

The effectiveness of these cleaning methods is summarized in Table 7.3. In detail, the results were stated as follows (56):

Although alkaline Rochelle salt cannot completely remove all the corrosion compounds on marine archaeological silver coins, it is considered as an important process that should be done prior to any other cleaning methods. This has been suggested because the alkaline Rochelle salt technique is not harmful for this kind of coin.

Table 7.3 Effectiveness of cleaning methods (56).

Method	Effectiveness
Alkaline Rochelle salt	Partially effective
Alkaline dithionite reduction technique	Not good
Electrolytic reduction technique	Not good
Alkaline Rochelle salt combined with alkaline dithionite reduction technique	Effective
Alkaline Rochelle salt combined with electrolytic reduction technique	Effective

The electrolytic reduction technique, without any prior treatment with alkaline Rochelle salt, is an effective method for the removal of a corrosion layer, but it is not recommended for silver coins that are alloyed with copper, because a risk of plating a layer of element copper on the surface of the coin may emerge.

On the other hand, the combined methods, i.e., alkaline Rochelle salt combined with either an alkaline dithionite reduction technique or an electrolytic reduction technique, are effective for cleaning of marine archaeological silver coins.

The surfaces of the coins arrived at a good condition. They became sleek and homogeneous. Also, the writings and the inscriptions appeared clearly on the surface of the coins (56).

Laser Cleaning. Corroded archaeological coins from the Najran treasure were studied (57). This treasure was discovered at the archaeological site of Al-Okhdood in Najran city, Saudi Arabia. It dates back to the era before Islam.

The coins were found to be covered with thick layers of corrosion products, resulting from the degradation of the coins that occurred during their long-term burial. They were distorted so much that no detail of the original surface could be detected. The corrosion layer had a nearly composite structure, including metallic remains, mineralized, metallic, insoluble phases and products formed from the interaction between soil components and metal corrosion compounds, and also soil particles.

A Q-switched Nd:YAG laser at 1064 *nm* was used for cleaning. It was found that this is a very suitable technique for cleaning. It was found that 30 pulses were the best condition for the cleaning and reveal the surface of a silver coin with a thick layer of corrosion. In contrast, 10 pulses are the best condition for cleaning and reveal the surface of silver coins that have a thin layer of corrosion such as a previously mechanically cleaned coin.

The results showed that all elements of the corrosion products that covered the tested coins were removed from the coins after the laser cleaning procedure.

Also, this technique will preserve the coins against deterioration, as cleaning the coins avoids the reactivation of the corrosion processes (57).

7.2.4.4 *Laser Cleaning of a Bronze Bell*

The experimental results of a UV laser cleaning treatment on an outdoor bronze bell, dating from the second half of the year 600 have been reported (58).

The deterioration of a bronze exposed to sulfur dioxide and nitrogen dioxide produces sulfuric acid and nitrous acid as:

$$SO_2 + NO_2 + 2H_2O \rightarrow H_2SO_4 + HNO_3 \quad . \qquad (7.5)$$

Thus, the corrosion rate of a bronze material tends to increase by dissolving any passivating film and by forming a solution of hygroscopic metal sulfates.

Energy dispersive X-ray fluorescence nondestructive analysis was performed on the bell before, during and after the cleaning treatment in order to assess the laser ablation threshold, to define the efficiency of laser cleaning process, to avoid possible damage of laser on the bell and to determine the concentration of the constituents of both the patina and the alloy.

The energy dispersive X-ray fluorescence portable apparatus was used in order to evaluate the variation of concentration of sulfur, chlorine, calcium, copper, lead and tin in the course of the laser cleaning procedure. The variation of the elements in the course of the laser cleaning is shown in Table 7.4.

The data indicate that in the course of laser cleaning treatments a gradual reduction and controllable concentration of sulfur and calcium occurs. Moreover, it has been demonstrated that the removal

Table 7.4 Variation of the elements in the course of the laser cleaning (58).

	Not treated	Ablation 1	2	3	4
Total counts	0	20	40	100	190
Laser dose/ [Jcm^{-2}]	0	13.6	28	70	133
S% (w/w)	3.3	2.8	2.8	1.6	1.0
Ca% (w/w)	15.7	14.8	14.7	12.6	13.1
Cu% (w/w)	40	49	49	54	55
Pb% (w/w)	1.8	2.3	2.3	2.6	2.9
Sn% (w/w)	14.1	16.2	17.5	17.9	17.6

of these compounds involves an increase of the concentration of elements that characterize the alloy, such as copper, tin and lead (58).

References

1. S. Reich, G. Leitus, and S. Shalev, *New Journal of Physics*, Vol. 5, p. 99, 2003.
2. A. Doménech-Carbó, M.T. Doménech-Carbó, and M.A. Peiró-Ronda, *Analytical Chemistry*, Vol. 83, p. 5639, 2011.
3. A. Doménech-Carbó, M.T. Doménech-Carbó, M.A. Peiró-Ronda, I. Martínez-Lázaro, and J. Barrio-Martín, *Journal of Solid State Electrochemistry*, Vol. 16, p. 2349, 2012.
4. A. Doménech, *Analytical Methods*, Vol. 3, p. 2181, 2011.
5. F. Scholz, L. Nitschke, and G. Henrion, *Naturwissenschaften*, Vol. 76, p. 71, 1989.
6. A. Doménech-Carbó, M.T. Doménech-Carbó, M.A. Peiró-Ronda, and L. Osete-Cortina, *Archaeometry*, Vol. 53, p. 1193, 2011.
7. I.M. Villa, *Archaeological and Anthropological Sciences*, Vol. 1, p. 149, 2009.
8. I.M. Villa, *Lithos*, Vol. 55, p. 115, 2001.
9. S. Baron, C.G. Tamas, and C. Le Carlier, *Archaeometry*, Vol. 56, p. 665, 2014.
10. Y. Kahanov, D. Ashkenazi, D. Cvikel, S. Klein, R. Navri, and A. Stern, *Journal of Archaeological Science: Reports*, Vol. 2, p. 321, June 2015.
11. M.H. Klaproth, *Beiträge zur Chemischen Kenntnis der Mineralkörper*, Vol. 1, Decker, Berlin, 1795.
12. R.C. Fierascu, R.M. Ion, and I. Fierascu, *Instrumentation Science & Technology*, Vol. 43, p. 107, December 2014.
13. A. Doménech-Carbó, S. Capelo, J. Piquero, M.T. Doménech-Carbó, J. Barrio, A. Fuentes, and W. Al Sekhaneh, *Materials and Corrosion*, Vol. 67, p. 120, April 2015.
14. D.A. Scott, *Studies in Conservation*, Vol. 42, p. 93, 1997.
15. M.T.S. Nair, L. Guerrero, O.L. Arenas, and P.K. Nair, *Applied Surface Science*, Vol. 150, p. 143, August 1999.
16. D. Scott, *Copper and Bronze in Art: Corrosion, Colorants, Conservation*, Getty Conservation Institute, Los Angeles, 2002.
17. A. Doménech, M.T. Doménech-Carbó, T. Pasies, and M. del Carmen Bouzas, *Electroanalysis*, Vol. 24, p. 1945, 2012.
18. J.W. Spence, F.H. Haynie, F.W. Lipfert, S.D. Cramer, and L.G. McDonald, *Corrosion Science*, Vol. 48, p. 1009, December 1992.
19. S. Feliu, M. Morcillo, and S. Feliu, Jr., *Corrosion Science*, Vol. 34, p. 403, 1993.
20. S. Bhattacharjee, N. Roy, A.K. Dey, and M.K. Banerjee, *Corrosion Science*, Vol. 34, p. 573, 1993.
21. J. Kobus, *Materials and Corrosion*, Vol. 51, p. 104, 2000.
22. V. Lazic, F. Colao, R. Fantoni, and V. Spizzicchino, *Spectrochimica Acta Part B: Atomic Spectroscopy*, Vol. 60, p. 1014, August 2005.

23. F. Di Turo, N. Montoya, J. Piquero-Cilla, C. De Vito, F. Coletti, G. Favero, and A. Doménech-Carbó, *Analytica Chimica Acta*, Vol. 955, p. 36, February 2017.
24. A. Doménech-Carbó, J.M. del Hoyo-Meléndez, M.T. Doménech-Carbó, and J. Piquero-Cilla, *Microchemical Journal*, Vol. 130, p. 47, 2017.
25. C.E. Bottaini, A.L.M. Silva, D.S. Covita, L.M. Moutinho, and J.F.C.A. Veloso, *X-Ray Spectrometry*, Vol. 41, p. 144, March 2012.
26. H. Coghlan, *Oxford: Pitt Rivers Museum Occasional Papers on Technology*, Vol. 4, p. 38, 1975.
27. P.T. Craddock and N.D. Meeks, *Archaeometry*, Vol. 29, p. 187, 1987.
28. T. Dilo, N. Civici, F. Stamati, O. Cakaj, and A. Angelopoulos, Archaeometallurgical characterization of some ancient copper and bronze artifacts from Albania, in A. Angelopoulos and T. Fildisis, eds., *AIP Conference Proceedings*, Vol. 1203, pp. 985–990. 7th International Conference of the Balkan Physical Union, American Institute of Physics, 2009.
29. I. Cacciari and S. Siano, *Journal of Infrared, Millimeter, and Terahertz Waves*, Vol. 38, p. 1, 2017.
30. S. Krimi, J. Klier, J. Jonuscheit, G. von Freymann, R. Urbansky, and R. Beigang, *Applied Physics Letters*, Vol. 109, p. 021105, July 2016.
31. M.J. Navas, A.G. Asuero, and A.M. Jiménez, *Applied Spectroscopy*, Vol. 70, p. 207, Jan 2016.
32. T. Palomar, B.R. Barat, E. García, and E. Cano, *Journal of Cultural Heritage*, Vol. 17, p. 20, 2016.
33. J.F. Asmus, *Technol. Conserv.*, Vol. 3, p. 14, 1978.
34. R. Pini, S. Siano, R. Salimbeni, M. Pasquinucci, and M. Miccio, *Journal of Cultural Heritage*, Vol. 1, Supplement 1, p. S129, 2000.
35. D. Jiang, Y. Luo, and M. Gao, Research with pulse laser to remove the rust on bronze, in *Proceedings of the EEC China Workshop on Preservation of Cultural Heritages. Xian, Shaanxi, PR Of China, September 25-30, 1991*, pp. 102–109. Teti, 1992.
36. C.A. Cottam, D.C. Emmony, J. Larson, and S. Newman, *Laser cleaning of metals at infra-red wavelengths* in W. Kautek and E. König, eds., *Lasers in the Conservation of Artworks*, Vol. 1, pp. 95–98. Mayer & Comp., Vienna, 1997.
37. C.A. Cottam and D.C. Emmony, *Practical assessment of the cleaning effect of TEA CO_2 laser radiation on corroded metal samples* in W. Kautek and E. König, eds., *Lasers in the Conservation of Artworks*, Vol. 2. Mayer & Comp., Vienna, 1999.
38. Wikipedia, Q-switching — wikipedia, the free encyclopedia, 2017. [Online; accessed 4-January-2017].

39. E. Angelini, S. Grassini, M. Olivero, M. Parvis, and G. Perrone, Laser cleaning of metal artifacts: Microstructural, chemical and optical fiber-based analysis, in *2016 IEEE International Instrumentation and Measurement Technology Conference Proceedings*, pp. 1–5, May 2016.
40. T. Palomar, M. Oujja, I. Llorente, B. Ramírez Barat, M.V. Cañamares, E. Cano, and M. Castillejo, *Applied Surface Science*, Vol. 387, p. 118, November 2016.
41. R. Cesareo, A. Bustamante, J. Fabian, C. Calza, M. Dos Anjos, R.T. Lopes, W. Alva, L. Chero, M. Espinoza, R. Gutierrez, R. Rodriguez, and M. Seclen, *X-Ray Spectrometry*, Vol. 40, p. 37, 2011.
42. A. Galli, L. Bonizzoni, E. Sibilia, and M. Martini, *X-Ray Spectrometry*, Vol. 40, p. 74, 2011.
43. E.A.O. Saettone, J.A.S. da Matta, W. Alva, J.F.O. Chubaci, M.C.A. Fantini, R.M.O. Galvão, P. Kiyohara, and M.H. Tabacniks, *Journal of Physics D: Applied Physics*, Vol. 36, p. 842, March 2003.
44. Wikipedia, Moche culture — wikipedia, the free encyclopedia, 2017.
45. R. Cesareo, A. Bustamante, J. Fabian, S. Zambrano, W. Alva, L. Chero, M. del Carmen Espinoza, R. Rodriguez, M. Seclen, F. Gutierrez, et al., *Journal of Materials Science and Engineering. B*, Vol. 1, p. 48, 2011.
46. Wikipedia, K-alpha — wikipedia, the free encyclopedia, 2016. [Online; accessed 14-September-2016].
47. E.F. Shaykhutdinova, A.G. Sitdikov, and R.K. Khramchenkova, *International Journal of Humanities and Cultural Studies*, Vol. 1, p. 14, 2016.
48. D.L. Hamilton, *Texas A&M University*, 1999.
49. D.L. Hamilton, Methods of conserving archaeological material from underwater sites, electronic http://nautarch.tamu.edu/CRL/conservationmanual/index.htm, 2010. Revison No. 2.
50. O. Mircea, I. Sandu, V. Vasilache, and A.V. Sandu, *Microscopy Research and Technique*, Vol. 75, p. 1467, 2012.
51. Wikipedia, Liesegang rings (geology) — wikipedia, the free encyclopedia, 2015.
52. E. Figueiredo, M.A.S. Pereira, F. Lopes, J.G. Marques, J.P. Santos, M.F. Araújo, R.J.C. Silva, and J.C. Senna-Martinez, *Spectrochimica Acta Part B: Atomic Spectroscopy*, Vol. 122, p. 15, August 2016.
53. I. Sandu, C. Marutoiu, I.G. Sandu, A. Alexandru, and A. Sandu, *Acta Universitatis Cibiniensis Seria F Chemia*, Vol. 9, p. 39, 2006.
54. I. Sandu, N. Ursulescu, I.G. Sandu, O. Bounegru, I.C.A. Sandu, and A. Alexandru, *Corrosion Engineering, Science and Technology*, Vol. 43, p. 256, September 2008.
55. R. Reale, S.H. Plattner, G. Guida, M. Sammartino, and G. Visco, *Chemistry Central Journal*, Vol. 6, p. S9, 2012.
56. O. Abdel-Kareem, A. Al-Zahrani, and A. Al-Sadoun, *Mediterranean Archaeology and Archaeometry*, Vol. 16, p. 107, 2016.

57. O. Abdel-Kareem, A. Al-Zahrani, A. Khedr, and M.A. Harith, *Mediterranean Archaeology and Archaeometry*, Vol. 16, p. 135, 2016.
58. G. Buccolieri, V. Nassisi, A. Buccolieri, F. Vona, and A. Castellano, *Applied Surface Science*, Vol. 272, p. 55, May 2013.

Index

Acronyms

Chemicals

Boldface numbers refer to Figures

General Index

Also of Interest

Check out these other books by the author published by Scrivener Publishing

Chemicals and Methods for Conservation and Restoration
Paintings, Textiles, Fossils, Wood, Stones, Metals, and Glass
By Johannes Karl Fink
Published 2017. ISBN 978-1-119-41824-5

Additives for High Performance Applications Chemistry and Applications
By Johannes Karl Fink
Published 2017. ISBN 978-1-119-36361-3

Metallized and Magnetic Polymers
By Johannes Karl Fink
Published 2016. ISBN: 9781119242321

Marine, Waterborne, and Water-Resistant Polymers Chemistry and Applications
By Johannes Karl Fink
Published 2015. ISBN 978-1-119-018486-7

The Chemistry of Printing Inks and Their Electronics and Medical Applications
By Johannes Karl Fink
Published 2015. ISBN 978-1-119-04130-6

The Chemistry of Bio-based Polymers
By Johannes Karl Fink
Published 2014. ISBN 978-1-118-83725-2

Polymeric Sensors and Actuators
By Johannes Karl Fink
Published 2012. ISBN 978-1-118-41408-8

Handbook of Engineering and Specialty Thermoplastics
Part 1: Polyolefins and Styrenics
By Johannes Karl Fink
Published 2010. ISBN 978-0-470-62483-5

Handbook of Engineering and Specialty Thermoplastics
Part 2: Water Soluble Polymers
By Johannes Karl Fink
Published 2011. ISBN 978-1-118-06275-3

A Concise Introduction to Additives for Thermoplastic Polymers
by Johannes Karl Fink.
Published 2010. ISBN 978-0-470-60955-2

Printed in the USA/Agawam, MA
September 16, 2026

910661.008